RISK INTELLIGENT SUPPLY CHAINS

How Leading Turkish Companies Thrive in the Age of Fragility

Praise for *Risk Intelligent Supply Chains*

"This is an amazingly timely and lucidly written book focusing on a topic of extreme importance. Even though this book is dealing primarily with Turkish industries, it is a must-read for supply chain specialists anywhere in the world."

—**Michael L. Pinedo,** *Julius Schlesinger Professor of Operations Management, Leonard N. Stern School of Business, New York University*

"Like a great classical composer, Professor Haksöz masterfully scores the four necessary voices of evermore critical *risk intelligent supply chains* (RISC) through his innovative and actionable I-Quartet model—Integrator, Inquirer, Improviser, and Ingenious. Supply chain executives, take note!"

—**Brian W. Hagen, Ph.D.,** *Managing Director of Decision Empowerment Institute*

"Through a number of fascinating case studies of some of Turkey's most successful companies, this book offers important new insights into how to best manage supply chain risk in these times of global uncertainty."

—**Fatih Birol, Ph.D.,** *Chief Economist and Director of Global Energy Economics, International Energy Agency*

"Çağrı Haksöz has written an important and original book. *Risk Intelligent Supply Chains* is a pioneering study that clearly elucidates the interrelated issues of risk and supply chains. In this timely and well-written book, Haksöz competently shows how leading Turkish companies manage risk in their supply chains. Çağrı Haksöz combines a deep understanding of supply chains and Turkish industry to produce an outstanding book that illuminates both theoretical and implementation questions applicable to a wide cross section of industries and companies worldwide."

—**Ömer S. Benli,** *Associate Dean, College of Business Administration, California State University, Long Beach*

"This is an excellent and thought-provoking book. Çağrı makes a compelling case for why top management should proactively manage supply chain risks to survive in a global and intensely competitive market. The operationalizing of the concept of *risk intelligent supply chain* is very clear and rational and provides a framework that managers can use to develop supply chain risk management capabilities. The case studies of Turkish companies provide actionable takeaways in managing supply chain risk. It is a must read for practitioners and academicians."

—**Vinod Singhal,** *Brady Family Professor of Operations Management, Scheller College of Business, Georgia Institute of Technology*

"*Risk Intelligent Supply Chains* is a thoughtful and timely book dealing with the topic of supply chain risk management in the context of emerging markets, interspersed with compelling data, metaphors, examples, and case studies. The strategic framework proposed is thought provoking and worthy of reflection by supply chain strategists. In presenting his thesis and field research, Çağrı Haksöz tells us insightful stories about managing risk in supply chains in emerging markets. The case studies and ideas presented in this book will be useful not only to managers and business students in Turkey but also to those in other emerging markets and MNCs seeking to enter such markets."

—**Vishal Gaur,** *Professor of Operations Management, The Johnson School, Cornell University*

"Professor Haksöz has made a strong contribution to the study and practice of risk management. He presents a coherent framework for approaching the myriad risks faced by enterprises competing in an increasingly interdependent global marketplace. The analogies he draws between adaptive supply chains and natural ecosystems display his profound understanding of these complex processes. The connections to art, music, history, and philosophy are similarly insightful. Finally, the examples of I-Quartet in practice at leading Turkish companies provide concrete context for application of his theoretical model. *Risk Intelligent Supply Chains* is an excellent resource for students, academics, and practitioners."

—**Seth Hoffman,** *Vice President and Chief Compliance Officer, American Investment Services, Inc.*

"Supply chain management is one of the key concepts that improve the competitive advantage of a company substantially. Moreover, it is one of the most critical areas in management that determine the success and failure of a company in today's highly competitive and volatile environment wherein information moves very fast. Companies have to find ways to improve customer services while at the same time bring down costs. Corporate management also needs to work on risk management issues throughout the company involving all the operational processes and financial issues. This book introduces the Risk Intelligent Supply Chain model (I-Quartet model), which I found very applicable and useful. The model is explained in detail throughout the book, and the various roles that a risk intelligent supply chain plays under different conditions and their interrelations are examined. These four roles are Integrator, Inquirer, Improviser, and Ingenious and are explained in the book in a very innovative and practical way. The last part of the book consists of case studies from Turkish companies. This part is written in an interview format, which improves the clarity of the concepts. Part 3 also helps readers familiarize themselves with the various industries covered in the book. I recommend the book to all practitioners."

—Zafer Kurtul, CFA, *CEO and Board Member, Sabancı Holding, Turkey*

"There are lots of books and proceedings on supply chains. But this book is different in the sense that it brings the concepts of fragility and resilience to the attention of the scientific community. It answers the question of how to become a risk-aware supply chain. I really enjoyed reading this book and strongly recommend it for everyone working in this field."

—İhsan Sabuncuoğlu, *Professor and Chair, Industrial Engineering, Bilkent University, Turkey*

"While global corporations have been snatching Turkish executives to tap into their expertise in dealing with geostrategic fragility, Professor Haksöz finds a way to analyze their innovative techniques to develop a unique model for risk intelligence. As an executive managing U.S., Turkish, and Indian operations, I find the I-Quartet model appealing to emerging as well as developed markets."

—İlhan Bağören, *CEO, Telenity, Inc.*

"Coming from an operational and quantitative background, I found Çağrı Haksöz's original look at strategic supply chain risk management revealing, informative, and full of actionable lessons. *Risk Intelligent Supply Chains* with its Turkish backdrop is captured vividly and skillfully, bringing together such disparate and diverse areas as history, art, nature, politics, literature, music (my favorite), and, of course, business. Breaking away from the books that focus on one thing at a time, Haksöz's observant mind and his cross-cultural upbringing produce a creative way of integrating ideas, research, teaching, and arguably more importantly, practice. Reading the refreshing, Charlie Rose-style conversations with executives, I no longer wonder why the Turkish economy is thriving at a time and in a region where crisis and stagnation are the norm, showing the underpinnings of the good old method of achieving prosperity, one company, one sector, one citizen at a time. In fact, I would argue that Haksöz's book is a counterexample to the impossibility of teaching wisdom: he achieves the impossible masterfully by distilling the wisdom of risk intelligent supply chains for the rest of us!"

—**Erhan Kutanoğlu,** *Associate Professor of Operations Research and Industrial Engineering, University of Texas at Austin*

"Haksöz's book is a good reference for risks in supply chains and addressing these in real life. The book has two main sections: theory and practice. The practice section has the refreshing approach of presenting interviews with supply chain executives and thereby provides us with unfiltered information that cannot be found anywhere else. This approach and information differentiate the book from others. The book focuses on supply chains of companies headquartered in Turkey. Hence, it is relevant for professionals working at or with Turkish companies or in other emerging markets."

—**Metin Çakanyıldırım,** *Associate Professor of Operations Management, Navin Jindal School of Management, The University of Texas at Dallas*

"In a quickly getting smaller world with intense and complex relations among companies and suppliers and partners, the largest risk is not to have the intelligence of risk. To understand all dimensions of risks in a complex supply chain and to manage them—while taking care of financial, physical, and geographical conditions—now need the *Risk Intelligent Supply Chains* (RISC) mind-set. In the presence of intense data flow, today's world now has more unknown and unknowable risks because of global warming, long-term and wide-based financial crises, wars, etc. More capacity is needed to understand and interpret this data flow and take the necessary steps to manage global risks. Based on Professor Haksöz's observations of natural ecosystems starting from childhood and his belief in the historical upbringing of Turkish people—honed during migratory journeys from Central Asia to Turkey—demonstrate how Turkish companies understand and manage risks to create the sixteenth largest economy of the world. This gripping book should be kept under the hand of every manager."

—**Ender Yılmaz,** *Treasury Operations Vice President Akbank, Turkey*

RISK INTELLIGENT SUPPLY CHAINS

How Leading Turkish Companies Thrive in the Age of Fragility

ÇAĞRI HAKSÖZ

CRC Press
Taylor & Francis Group
Boca Raton London New York

CRC Press is an imprint of the
Taylor & Francis Group, an **informa** business

CRC Press
Taylor & Francis Group
6000 Broken Sound Parkway NW, Suite 300
Boca Raton, FL 33487-2742

Printed on acid-free paper
Version Date: 20130214
International Standard Book Number-13: 978-1-4665-0447-9 (Hardcover)
Cover art by Professor Yankı Yazgan

Library of Congress Cataloging-in-Publication Data

Catalog record is available from the Library of Congress

Visit the Taylor & Francis Web site at
http://www.taylorandfrancis.com

and the CRC Press Web site at
http://www.crcpress.com

To my grandparents,
Zernişan and Mustafa Haksöz
Munise and Ali Osman Alırsatar

“Seek and find…”
Hacı Bektaş-ı Veli

“The heart speaks as it becomes silent…”
Mevlana Celaleddin Rumi

“Once love arrives, every shortcoming quits…”
Yunus Emre

Contents

Foreword

I congratulate Çağrı Haksöz for writing a much needed synthesis of theories on managing supply chain risk. Much has been written to praise the benefits of globalization. It also has its dark side. Reading many examples in *Risk Intelligent Supply Chains* itself reveals how widespread, complex, exciting, and vulnerable our supply networks have become. Single incidents at factories, supply sources or pathways can jam the commercial highways of the world. Even the electronic Web around us is highly vulnerable to disruptions and failures. The best run banks, global conglomerates, factories, and universities have experienced disruptions that are unparalleled in recent history. There is a growing feeling that firms and organizations in general should be better equipped to manage risk, perhaps by anticipating, forecasting through story telling, training, or whatever it takes to equip and prepare for the failures.

When I teach a product recall case, such as the *Ford-Firestone* case that I wrote with Michael Pinedo and Eitan Zemel after the tire crisis, I ask why is it that everyone sounds so intelligent after a fault or problem is uncovered? Where were these intelligent people when the sequence of events was unfolding? Why are we so sure after the fact? How do we continue to second guess even the best and the brightest? Haksöz provides many important reasons why even the best decision-makers and planners can be led astray, so often, and perhaps in the same way. To top this we now accept that mutations can occur and random combinations can lead to extreme behavior of people and systems, that is, the unexpected. Haksöz provides ways of anticipating the unexpected.

Combining theory, practice, and experience gained through years of consulting, teaching, and research, Çağrı Haksöz proposes a framework for examining and managing risk which he calls the *I-Quartet Model.* He proposes four important roles that contribute to the supply intelligence, the roles of the *integrator, inquirer, improviser,* and the *ingenious.* The framework can be used to systematically diagnose the risk intelligence of a supply chain. It provides a starting point for a dialogue on risk management in supply chains.

Haksöz illustrates the framework with three insightful case studies. Each case study illustrates the use of the framework. I am sure every manager will enjoy and learn from applying these ideas to his or her supply chain network and by taking stock of how different roles and tasks in risk management contribute toward risk intelligence.

Sridhar Seshadri, Ph.D.
Department of Information, Risk, and Operations Management
McCombs School of Business, the University of Texas at Austin

Foreword

Risk, complexity, uncertainty, and fragility have become the norm for today's business world. New types of risks emerge every day in global supply chain networks that require fresh thinking and execution. Understanding the dynamics of risks as well as their interactions is necessary to achieve resilience for individuals as well as companies, supply chains, and countries. As a result of global business environment megatrends that incorporate but are not limited to shorter product life cycles, transparency of information, and lessened profit margins, global competition has increased between supply chain networks, and hence the risk management capability of companies will differentiate leaders from laggards. According to the International Air Transport Association, the airline industry's overall profit margin for the past 10 years is negative.

In such a world, a growing number of Turkish companies manage not only to be resilient but also to thrive and create exemplars in supply chain risk management to other companies worldwide. To this end, with his unique book, Çağrı Haksöz contributes to the business world and advises on how to become a *risk intelligent supply chain* to be resilient and long lasting. There are at least five significant points regarding this book's originality and value that should be underlined.

1. It presents a thorough discussion on risk and resilience in global supply chain networks from very different perspectives such as biological ecosystems, complexity theory, near-miss management, extreme value and game theory, operational risk, and musical improvisation and creativity.
2. It introduces the risk intelligent supply chain concept for the first time to the business and academic world.
3. It proposes a novel model (I-Quartet model) that aims to help business executives on the journey toward a risk intelligent supply chain.
4. It acts as a pioneer, and presents unique and interesting practices and strategies from leading Turkish companies—some with dynamic conversations (a novel method in itself) to create a collegial atmosphere for thinking and learning together.

5. It opens new paths for exploration, research, and field studies in supply chain risk management and risk intelligence in emerging as well as developed economies.

In short, Haksöz has written a very interesting book of lasting value by bringing his unique style of telling stories and out-of-the-box metaphorical thinking. I am sure any business professional around the globe as well as a management scholar who is interested in the intricate details of risk intelligence in the context of global supply chain networks will benefit from what is being offered. My only hope and wish are that Çağrı Haksöz continues to produce and share such creative works in the years to come.

Sami Alan

Senior Vice President for the Regional Flights (AnadoluJet) Turkish Airlines

Prologue: A Tale of Fragility and Resilience

> Life is reconciliation of opposites. It is death if war erupts between them.
>
> **Mevlana Celaleddin Rumi**[1]

> There is no subject so old that something new cannot be said about it.
>
> **Fyodor M. Dostoyevsky**[2]

Welcome to our world of *fragility* and *resilience*. As the world becomes interdependent and tightly coupled in terms of organizations, supply chain networks, industries, countries, as well as individuals, information, and cash, fragility is imminent. Look at the economic and financial crises in the United States, Europe, and elsewhere. Interestingly, the mirror image of fragility is resilience. An individual, organization, supply chain network, industry, or country can continue to exist, however, only by acknowledging the dual presence of fragility and resilience and then by reconciling the two.

To reconcile fragility and resilience for supply chain networks, I set out on my journey to tell the story of *risk intelligent supply chains*, which I define as the supply chains that thrive and are resilient in the age of fragility. Even though the telling of the story is recent, the tale itself has been living with me for many years. Actually, this story is a culmination of ideas, concepts, and models I have been thinking about and working on since my graduate studies at New York University's (NYU's) Stern School of Business. Some of these ideas were published in various forms. Yet, as years went along, a book seemed necessary to structure and present these ideas and concepts in a coherent whole. Hence, this book was born with this intention and motivation.

This book becomes my first step on the ladder to understand, think, and reflect on supply chain networks and how they succeed or fail. Because fragility and resilience cannot exist independently, success and failure also coexist. They live together in the minds of individuals, organizations, and supply chain networks. Surely, sustainable and perpetual success is what is being sought. Yet what we have been observing in many organizations and supply chain networks is that success is ephemeral and very difficult to sustain. After a while, frailties appear and, if not correctly attended to and managed, can lead to eventual fragility and perhaps death.

In writing this book, my research, teaching, and consulting on supply chain risk for many years formed the essential background on which I could build the frames of my work. However, I need to mention *three major events* in my life that crucially affected my vision and thinking on ecosystems, resilience, fragility, and risk.

The very first event occurred during my childhood when I spent eight years in the lovely city of Afyonkarahisar in Turkey. I had lived in the suburbs of the city where sparrows, turtles, honeybees, butterflies, and many more lived in an ecosystem surrounded by pine trees and green meadows. During that period, my one-to-one interaction with many animals and plants had sown a seed of love in my heart for nature. The second event occurred around twenty years ago in Ankara, Turkey. I was lucky enough to study in the pioneering science high school of Turkey—Ankara Science High School (AFL)—under the mentoring of extraordinary teachers. In that school, my admiration for nature and its various beauties led me to study biology in more detail, eventually making me a Biology Olympiads medalist. During that period, I learned the intricate details of the organisms, ecosystems, and nature about which I had wondered in my childhood. Over the years, my liking and interest for the natural world have dramatically increased as I have wandered different cities and countries. The third event occurred in Banff, Alberta, Canada. In May 2004, when I visited Banff National Park for a conference, I was mesmerized by the pristine nature of the Canadian Rockies, its turquoise lakes, amazing trees, and animals (elks, bald eagles, moose, grizzly bears, and especially the gray wolves and their cubs). How these very different habitats interacted and lived together with each other peacefully as well as how these ecosystems continued their livelihood in resilience created a major epiphany in my perception of risk and fragility.

As the years passed, I became a careful observer of the natural ecosystems and their resilience under harsh conditions. One day, when working on my Ph.D. at NYU Stern, my major research question emerged: How can an organization and a supply chain network emulate a natural ecosystem to achieve sustainable resilience and risk intelligence? I had been contemplating this question for some time until I returned to Turkey to teach at the Sabancı School of Management in İstanbul. Since my arrival in the city, I have been teaching and advising Turkish executives who are highly adaptive and creative. During this period, I had the luck of meeting the brightest minds in various sectors, who were leading their companies as well as supply chains much better than the rest of the market. It turns out that these leading Turkish companies have developed original ways of managing risks in their supply chain networks. So I decided to study their novel practices and then encapsulate their strategic insights into actionable models for others to emulate. In this long and arduous journey, I had to stop for a time, rest a while, and rejuvenate just like a caravan rests in a caravanserai on the Silk Road. Thus, I consider this book to be my first caravanserai in which I tell the story of risk intelligent supply chains and present a number of insightful lessons obtained during this leg of the journey.

This book is unique and somewhat avant-garde for a number of reasons. First, it brings the *risk intelligent supply chain* (RISC) concept to life for the first time. The concept of *risk intelligence* per se does not belong to me. It has recently existed in practice with various interpretations.[3] However, it is the first time that the *risk intelligent supply chain* concept, together with its strategic and operational characteristics, is being discussed in book length. Second, the book is also avant-garde in the sense that a diverse set of disciplines besides its core—supply chain and risk management—is brought together in a unified whole via a pragmatic and insightful risk intelligent supply chain concept. Third, the book presents never before published cases and practices of leading Turkish companies that thrive globally with their supply chain risk intelligence.

The book consists of three parts: (1) Evolving Global Supply Chain Networks: Fragile, Capricious, and Entangled; (2) How to Become a Risk Intelligent Supply Chain; and (3) How Do Leading Turkish Companies Thrive in the Age of Fragility?

In Part 1, I set out to demonstrate the increasing variety, frequency, and impact of risks in global supply chain networks. I show that internal and external disruptions of all kinds adversely affect global supply chain networks. Interdependencies and complex interactions exacerbate this situation. To this end, this part aims to open our eyes and ears to prepare us for the upcoming chapters. It clearly shows that there is an immediate need for supply chain networks at all scales (temporal as well as spatial) to be more risk intelligent. Yet it does not show how.

Part 2 presents the details of a risk intelligent supply chain by proposing the *I-Quartet model*. In short, it answers the question of how to become a risk intelligent supply chain. In the conceptual model proposed, there are four essential roles—*integrator, inquirer, improviser,* and *ingenious*—that any supply chain network must play to become risk intelligent. This part consists of five chapters. Beginning with Chapter 2, which provides a brief overview of the I-Quartet model, each chapter introduces the capabilities, skills, characteristics, and wisdom for each I-Quartet role in turn.

In Part 2, you will notice the confluence of diversity of ideas, concepts, methods, and techniques from a set of very different scientific and management disciplines. This method of synthesis was necessary to move beyond descriptions and travel toward the heart of the risk intelligent supply chains. To this end, I aimed to amalgamate ideas and concepts from biology and ecosystems (resilience, evolution, animal and plant behavior), social psychology, medicine, Silk Road and Turkish art history, enterprise and operational risk management, financial risk hedging, financial derivatives and real options, risk perception and biases, complexity and complex adaptive systems, fractals, self-organization, networks, environmental sciences, earthquake and epilepsy modeling, extreme value theory, game theory, organizational behavior, business strategy, creativity and innovation, both musical (jazz and Turkish music) and organizational improvisation, normal accidents theory, high reliability management, and safety/accident/near-miss management. Interested readers can go beyond what is discussed in this book and consult the original sources available in the extensive Bibliography. In Part 2, every I-Quartet role chapter concludes not with directives but with strategic questions that every executive should ask about and reflect on in the journey toward a risk intelligent supply chain.

In Part 3, after a prelude on transmitting wisdom of companies to others, I present three interesting cases of leading Turkish companies—Kordsa Global, Brisa, and AnadoluJet—which vividly demonstrate the idiosyncratic strategies, practices, and wisdom of managing risk intelligent supply chains in an emerging market. I choose to present these cases because Kordsa Global and Brisa comprise the critical global upstream players of the *global tire supply chain network,* whereas AnadoluJet is a second-tier customer at the downstream of the same supply chain. The Kordsa Global and Brisa chapters are presented in dynamic conversations. However, AnadoluJet suited an essay format so that the complete story from inception to actual operations could be told more succinctly.

In this book, readers will notice that I use a variety of *metaphors* inspired by music, nature, art, sports, and health care. The first fundamental metaphor is coming from the classical music scene. I love the pieces performed by string quartets. Using a string quartet analogy, I created the I-Quartet model, a quartet of four essential roles every risk intelligent supply chain must play harmoniously. This model forms the theoretical underpinning of risk intelligent supply chains. The second fundamental metaphor appearing in Part 2 of the book is called the *Acrobat over the Bosphorus.*[4] Beginning with the book cover art, where an acrobat on a tightrope crosses the Bosphorus in İstanbul, each I-Quartet role is visualized with this acrobat metaphor. The third fundamental metaphor that makes up the infrastructure of a risk intelligent supply chain is *Swan Lake,* with white, gray, and black swans. The Swan Lake metaphor is inspired by the Black Swan concept in risk, yet it has been expanded and embellished much further. A Swan Lake represents the lake of risks of various sizes and likelihoods that appear in global supply chain networks. In addition to these two metaphors, the book includes other metaphors such as *Turkish Carpet Weaving, Shivers Model, Ladder of Mastery,* and *Ocean Exploration.* The main reason I use these metaphors is that it makes conveying the insights of my messages to business professionals much easier. I tried and tested this method with success while consulting on projects and teaching executives. So I do hope that these metaphors increase the yield in the gardens of your creativity and enable better ideas to bloom.

Let me say a few words on why Turkish companies have been chosen for a globally important concept of risk intelligent supply chains. After proposing the conceptual underpinnings of a risk intelligent supply chain in Part 2, I present leading Turkish company examples. First, such cases from Turkey have not yet appeared in management literature. To this end, this book is a pioneer that lays the groundwork for future studies in Turkey as well as other emerging and developed economies. Surely, I could have very well used case examples to demonstrate risk intelligent supply chain insights of non-Turkish companies. In fact, there are many highly successful firms in other emerging markets such as China, India, Brazil, Russia, and Mexico that manage their risk intelligent supply chains effectively.

This book aims to demonstrate that a number of Turkish companies, some of which are presented here, epitomize organizations that create and lead risk intelligent supply chains. Cases from Turkey are significant because, in light of their location in a geostrategically fragile region, Turkish companies need to intelligently take risks and orchestrate their supply chain networks while continuously improvising and learning from their failures and weak signals. To this end, I think that strategic insights and especially the wisdom behind those insights are too important to be overlooked and lost in the frenetic business life. Hence, I decided to share these insights of Turkish executives with the global business community. A major caveat is that these insights should be evaluated in their particular contexts. If these insights do in fact inspire readers from around the world, motivate and energize them to take actions to begin a journey toward a risk intelligent supply chain, I consider my goal achieved.

All in all, my firm belief is that today risk intelligent supply chains are necessary to lead and thrive in the midst of the complexity and fragility. The road to becoming a risk intelligent supply chain needs a Ladder of Mastery, which is an ever-continuing journey with a caravan. This book is a story of those companies that climb on this ladder as they travel with a caravan each day. And it is also my personal story dedicated to understanding, learning, and explaining the intricate details and beautiful images of this caravan travel. I know that I have reached the very first step of this ladder in this book—the very first caravanserai. There is more to be explored and more paths to be traveled. Nevertheless, Ralph Waldo Emerson whispers into my ears and somewhat consoles me: "The reward of a thing well done is to have done it." So, why wait? Open wide your minds and hearts so we may travel together with this caravan exploring risk intelligent supply chains and learning from leading Turkish companies.

As a final word, I am interested in your comments, thoughts, and examples on the ideas, concepts, and arguments presented in this book. For sharing your ideas, please reach me via email (cagrihaksoz@sabanciuniv.edu and cagri.haksoz@gmail.com), or alternatively post comments on Twitter with the following hashtags:

Part 1:
- #RISC
- #RISCFragility
- #RISCResilience
- #RISCFragility+Resilience

Part 2:
#RISCIntegrator
#RISCInquirer
#RISCImproviser
#RISCIngenious
#RISCIQuartetModel
#RISCSwanLake
#RISCAcrobat
#RISCShiversModel

Part 3:
#RISCKordsaGlobal
#RISCBrisa
#RISCAnadolujet

Çağrı Haksöz
İstanbul, Turkey

Acknowledgments

A person is a cumulative sum of his past moments. I am no different. Living in six cities and three countries since my birth, I most certainly have enjoyed learning from many people of all ages and backgrounds. As I continuously learn, I strive to put something valuable on the table that will be appreciated by others. What you will read in this book is an amalgamation of that diverse set of lessons gained moment by moment. I am just the storyteller of the wisdom of my ancestors, great scholars and scientists, and highly innovative business people.

In the beginning, there were teachers and mentors. I cannot count them all. I am grateful to all my teachers at Dumlupınar Primary School (Afyonkarahisar), Çorum Anadolu Lisesi, and Ankara Science High School. All of them lit a candle on my path, especially Ali Altındiş and Abdulkadir Ozulu. My alma mater, Bilkent University, has a special place in my heart. I would like to thank its founder, İhsan Doğramacı, and all the world-class professors, especially İhsan Sabuncuoğlu and Ömer S. Benli, whose belief and support made a huge difference in my life. I am grateful to the dean of Sabancı School of Management at Sabancı University, Nakiye Boyacıgiller, as well as to the president of Sabancı University, Nihat Berker, who have both been working day and night to create a world-class institution. As the years moved along, I consider myself lucky to have met and worked with two giant scholars on Earth, Sridhar Seshadri of the University of Texas at Austin and Michael Pinedo of New York University. Their constant support and belief remain instrumental in my success today. I also thank all my professors and colleagues at NYU Stern, Cass Business School, NYU London, and Sabancı University, with whom I interacted for the past thirteen years. A number of individuals made contributions to this book in different ways throughout the years. My appreciation goes to Metin Çakanyıldırım, Erhan Kutanoğlu, Ashay Kadam, Koray D. Şimşek, Ali Devin Sezer, Halis Sak, Vishal Gaur, Ananth V. Iyer, David Juran, Christopher Tang, ManMohan Sodhi, Aaron Tenenbein, Aswath Damodaran, Vinod Singhal, Victor Araman, Paul Kleindorfer, Seth Hoffman, Ganesh Janakiraman, Melten Önder, Bülent Ünel, Mustafa Ünel, H. Mete Soner, Savaş Dayanik, Feyzullah Eğriboyun, N. Onur Bakır, Burcu Adivar, Turgut Öztürk, İsmail Cem Atalay, Murat Çokol, Aydın Albayrak, Murat Kaya, Gürdal Ertek, Ali Koşar, Tevhide Altekin, Ahmet

Öncü, Can Akkan, Semih Sezer, Nihat Kasap, Wojciech Machowiak, Kürşad U. Akpınar, Gülnur Muradoğlu, İbrahim Sirkeci, Berrak Dağ, M. Doğan Üçok, Murat Taş, OPIM group colleagues at SOM, and the International Supply Chain Risk Management network members.

We all know that business executives are time starved. I am grateful to a number of key individuals who were instrumental in this book such that these Turkish cases were brought to light. The generous support and contributions of especially Sami Alan, Bülent Bozdoğan, Arzu Öngün Ergene, Fatih Tunçbilek, and İdil Z. Taner Ertürk are greatly appreciated.

Research assistance of my students Birkan İçaçan, Alperen Manisalıgil, Arif Uygar Duran, Damla Durak Uşar, İlkan Sarigöl, Kübra Kandemir, Hazal Melis Camcı, and Rahman Saruhan is greatly acknowledged. Without their meticulous work, the book would not exist. I am also thankful to my students, from undergraduates to executives (as well as high school students at Sabancı University Summer School), at the NYU Stern School of Business; Cass Business School, London, UK; and Sabancı School of Management, Sabancı University, Turkey, who have taught me how much more I needed to learn. The very timely and diligent editorial assistance of Nancy Karabeyoğlu, Patrick Charles Lewis, Nilay Narmanlioğlu, and Alper Baysan at Sabancı University was invaluable. Their comments and suggestions made the text lively and readable. My editor, Lara Zoble, project coordinator Laurie Schlags, and project editor Karen Simon provided the necessary support and kept me on track. I am also thankful to my resourceful colleagues Işıl İnoğlu, Neyir Özdemir, Ülkü Köknel, Steve Moroz, and İpek İzet at Sabancı School of Management who made my academic life easier.

If you buy the book because you like the cover art, displaying an *Acrobat over the Bosphorus*, then you should meet Yankı Yazgan. Being one of the most respected child psychiatrists in Turkey, he took time in his busy schedule to create the original cover art as well as the insightful drawings in Chapters 3 to 6 that brought my metaphor to life. I am grateful to him as his contribution to this book is invaluable.

I want to thank my family, my father (Mehmet Haksöz), mother (Necla Haksöz), and sister (Burcu Haksöz), as they were always with me in any passage of my life as with this book. Their support, encouragement, belief, and love kept me going.

About the Author

Çağrı Haksöz is a faculty member of operations and supply chain management in the Sabancı School of Management at Sabancı University in İstanbul, Turkey, and an affiliate member of the Regent's Centre for Transnational Studies at Regent's College in London. He is the co-director of the MIT–Sabancı University Energy Security Initiative (MSESI). His current research focuses on supply chain risk management, risk intelligence, supply chain and energy contract management, product recall risk management, managing supply chain networks on the Silk Road, climate change and weather risk, and creativity and improvization. His previous books include *Managing Supply Chains on the Silk Road: Strategy, Performance, and Risk* (with Sridhar Seshadri and Ananth V. Iyer, 2011) as well as *Global Perspectives: Turkey* (with Metin Çakanyıldırım, 2012). He consults for national and international firms as well as government organizations.

1

EVOLVING GLOBAL SUPPLY CHAIN NETWORKS: FRAGILE, CAPRICIOUS, AND ENTANGLED

Chapter 1

Global Supply Chain Networks in the Age of Fragility

> In physics the truth is rarely perfectly clear, and that is certainly universally the case in human affairs. Hence, what is not surrounded by uncertainty cannot be the truth.
>
> **—Richard P. Feynman**[1]

> He who is afraid of the sparrow does not sow millet.
>
> **—Turkish Proverb**[2]

1.1 Welcome to the World of Extreme Risks and Opportunities

Today, if you ask anyone living in any city on the globe, you will hear the response that there is turbulence, ambiguity, complexity, and fragility around the world in firms, markets, countries, societies, and individuals. If you run *Google Books NGram Viewer*[3] for the key words "risk," "complexity," and "uncertainty" in the corpus books in English during 1800–2008, you obtain the data presented in the graph in Figure 1.1. It clearly shows that although the use of "complexity" and "uncertainty" is increasing, the occurrence of "risk" is increasing much faster, especially after the 1970s.

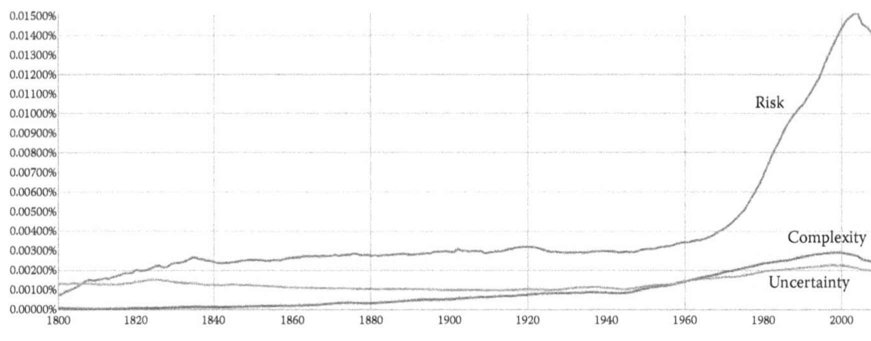

Figure 1.1 Google NGram Viewer results for "risk-complexity-uncertainty" in the *Corpus of English Books* (1800–2008).

If you are more interested in Google hits than printed words on paper, here are the results that might surprise some of you. The word "risk" offers 1,030 million hits, whereas the term "complexity" tops 111 million hits, and finally the term uncertainty is the least of three, getting only 88.8 million hits.[4] What do these results tell us?[5]

As fragility and uncertainty increase, global supply chain networks become more complex and the variety of risks behaves more interdependently. Cascading events are more frequent. Unprecedented consequences for the societies, corporations, and individuals are around the corner. Some of these events have already occurred—as seen in the Japanese Fukushima Disaster, the Thailand floods, and the financial crisis looming in Southern Europe. Of the 559 organizations surveyed by the Business Continuity Institute, for example, 85% experienced at least one supply chain disruption in 2011.[6]

Let us look at the cost of damages caused by these global disasters, as displayed in Figure 1.2. We observe a dramatic increase in damage cost and volatility in recent years.[7] During 1960–2011, global disasters that have caused the highest cost damages were *storms* (36%), *earthquakes* (31%), and *floods* (23%). The remaining costs were due to droughts, wildfires, extreme temperatures, and industrial accidents.

Energy is essential to run global supply chain networks at different stages within the context of the *energy–climate–water* trilogy. Yet energy has to be managed considering climate change impact. Emission-related risks are biting back the profits of supply chain networks. Fatih Birol, of the International Energy Agency, warned in March 2012:

> ... Economic concerns have diverted attention from energy policy and limited the means of intervention. As the CO_2 emissions rebounded to a record high level, by 2017, without further action, all CO_2 emissions will be locked in by the existing power plants, factories, and buildings. Unless bold and significant actions are taken, this will continue towards an undesirable regime for all of us.[8]

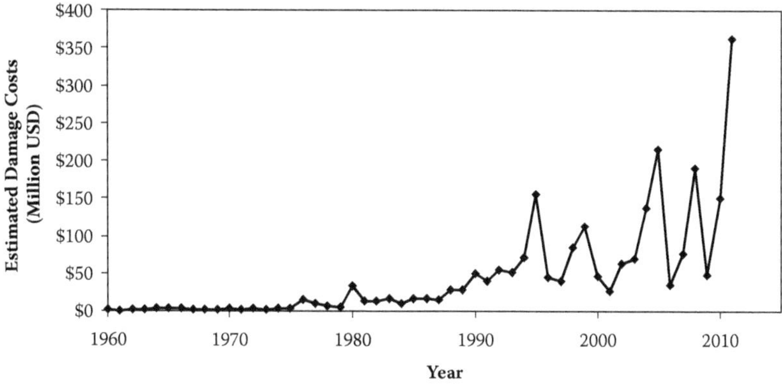

Figure 1.2 Estimated damage costs of global disasters, 1960–2011 (Million USD).

The *Global Risks 2011* report by the World Economic Forum has also emphasized that various global supply chain risks are increasing.[9] In the Global Risks Landscape 2012 map constructed with input from more than 450 experts, risks related to global supply chain networks emerged are shown in Table 1.1.[10]

When we focus on resource use, we observe that the extreme volatility of recent years in commodity and energy markets has caused executives to lose a good deal of sleep. A close look at Figure 1.3 where the *commodity food, metal,* and *energy price*

Table 1.1 Risks in Global Supply Chain Networks

• Water supply crises
• Food shortage crises
• Extreme volatility in energy and agriculture prices
• Critical systems failure
• Major systemic financial failure
• Failure of climate change adaptation
• Land and waterway use mismanagement
• Rising greenhouse gas emissions
• Species overexploitation
• Vulnerability to pandemics
• Mineral resource supply vulnerability
• Massive incident of data fraud and theft

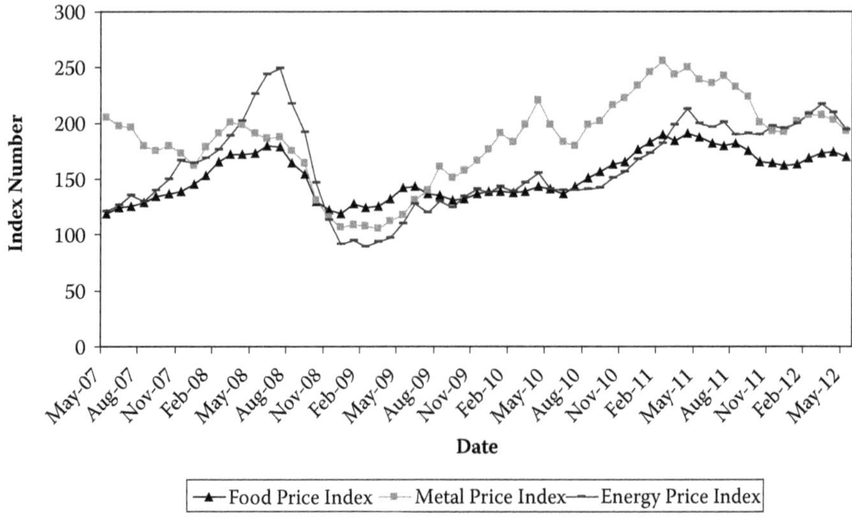

Figure 1.3 Commodity food, metal, energy price indices (May 2007 to May 2012; 2005 = 100). (From: http://www.indexmundi.com.)

indices are plotted during May 2007–2012 is sufficient to understand the potential impact of the volatile prices on global supply chain networks.[99]

On the other hand, fragility and complexity are not only due to natural disasters. There are man-made disasters that might also have extreme impacts. John Casti addresses some of the major catastrophic events, such as global pandemic, collapse of globalization, and the breakdown of the food supply, in his most recent book, *X-Events*. X-events are the extreme events caused by the collective actions and reactions of individuals. In fact, according to Casti, the "social mood" of human beings and societies is critical in anticipating these extreme events.[11] He sometimes draws a bleak picture but also offers hope.[12] He proposes the cause of these extreme events as follows:

> Complexity overload is the precipitating cause of X-events. That overload may show up as unmanageable stress and pressure in a single system, be it a society, a corporation, or even an individual. . . .[13]

The asynchrony of various risk-producing processes in the supply chain network creates an extra complexity; according to Alvin and Heidi Toffler. The authors devote a full section on timing and synchronization in their book *Revolutionary Wealth* and offer the following hypothesis:

> . . . The synchronization industry will expand because rising competition requires innovation after innovation, each of which, in turn,

> changes timing requirements and requires resynchronization. But the hidden paradox of the law of de-synchronization is that the more you synchronize at one level in a system, the more you de-synchronize at another.[14]

In global supply chain networks, the complexity of the system is caused by interactions of multiple stakeholders. There are independent and interdependent players in the game. For example, within a particular supply chain network, a firm can be simultaneously a competitor and a collaborator, that is, a *co-opetitor*, with another firm.[15] In such a vast network of suppliers, partners, collaborators, competitors, and customers, it is almost impossible to anticipate, understand, and interpret the behavior of these complex supply chain networks. Yet some people still believe in the value of accurate forecasting of uncertainty in the business world.[16] In fact, supply chain networks operating globally become tightly coupled dynamical systems that are susceptible to internal and external failures and disruptions.

The scale of interactions may also differ. Many different supply chain networks that operate in tandem can interact. For example, in automotive and toy supply chain networks, one of the key suppliers to both networks, plastics material producers, may cause extreme impacts if disrupted.[17]

All practitioners in the field around the globe agree that the fragility of supply chains has increased. A recent survey conducted by McKinsey & Company shows that 68% of global executives believe that the supply chain risk would increase in the following five years.[18] What is required is more resilience and robustness to prevent supply chain breakdowns in global supply chain networks.[19]

In addition to my explanations so far, the increasing variety and diversity of product and service offerings to a multitude of customers have exacerbated the complexity of global supply chain networks. For example, in 2009 the number of stock-keeping units (SKUs) in a North American grocer exceeded 100,000.[20]

Thus, it seems evident that supply chain risks in such a turbulent, fragile, and capricious world are evolving toward both the *unknown* and *unknowable*.[21] Ralph Gomory first proposed this classification of risks in 1995, proposing that there existed categories of known, unknown, and unknowable risks.[22] Later, U.S. Defense Secretary Donald Rumsfeld provided an aphorism based on the same idea and four uncertainty classes: *Known Knowns (KK), Known Unknowns (KU), Unknown Knowns (UK),* and *Unknown Unknowns (UU).* More recently, Casti offered an overview of these risks and the key methods to effectively manage them, based on the availability of data and model as shown in Table 1.2.

In this classification, *unknown unknowns* are the extreme events that can also be rare, possessing low probability and high impact (severity). These events have been identified as *black swan events* and were popularized by Nassim N. Taleb in his book *The Black Swan*.[23] Prior to the publication of this book, in 2005, a report by Deloitte Research emphasized the importance of the low-probability and high-impact losses and interdependencies in global supply chain disruptions.[24] So an

Table 1.2 Four Types of Uncertainty

	Data	No Data
Model	Known Knowns (Network analysis, dynamical systems theory)	Unknown Knowns (Simulation)
No Model	Known Unknowns (Statistical methods)	Unknown Unknowns (Scenarios, stories, creative imagination)

Source: Adapted from Casti.[87]

extreme event might occur at any time and disrupt the supply chain network, causing a loss of capital, talent, information, material, and customer.

The picture I have drawn up to now is gloomy, yet the story of fragility and complexity in global supply chain networks is only partially complete. Parallel to these undesirable changes over the globe exists opportunity and resilience for those who can see them. Risks carry the seeds of opportunities in themselves. In Chinese, risk has a dual meaning of danger and opportunity.

At this time in history, more types of data of any scale and scope can be collected via many internal and external sources such as intranets, enterprise resource planning and customer relationship management systems, business-to-business market exchanges, global financial and commodity markets, social media, and ubiquitous networks. The so-called Big Data phenomenon is currently sweeping countries, industries, and companies both large and small. This instant, diverse, and high-quality data revolution demands from supply chain managers much more effective supply chain risk management.

It is reported that in 15 of the 17 U.S. sectors, companies of more than 1,000 employees store, on average, 235 terabytes of data, more than the data stored by the U.S. Library of Congress. Any flow in the global supply chain networks, whether it is a financial transaction or a customer interaction, creates data.[25] Surely, collecting and capturing data is the first necessary step to create value. More importantly, interpreting, analyzing, and acting on data are essential to gain and sustain competitive advantage. Recent research has indicated that the productivity, profitability, and market value of companies positively correlate with the effective use of data and analytics.[26]

As opportunities are capitalized on, global success stories arise in various sectors, a phenomenon demonstrating that the management of supply chain risks may be indeed valuable. These stories provide for the remaining companies strong cases: for example, that intelligently managing supply chain risks is possible if a strong will, intention, and practice exist. For example, Hewlett-Packard has reported cost savings of around US$425 million by adopting a portfolio approach for managing supply chain contracts and related risks.[27]

In the fast fashion industry, the famous success story of Zara has shown that various quality-, timing-, and inventory-related risks can be effectively managed by designing, controlling, and managing an integrated global supply chain network.[28] The smooth integration of diverse flows such as material, product, information, and cash is necessary while managing supply chain risks.

In the following chapters of the book, we will see that hope, opportunities, and novel approaches still exist in the context of the global supply chain networks managed in Turkey by Turkish executives. These approaches create exemplars to be emulated for aspiring risk intelligent supply chains.

1.2 The Ubiquity of Global Supply Chain Risks

Considering all types of potential risks that may adversely affect the global supply chain networks, an exploratory study was conducted with the aim of deciphering the ubiquity and evolution of the global supply chain risks during 1999–2011. Global supply chain risks were classified into three major categories. As such, not only natural disasters and man-made disruptions, but also demand–supply mismatch risks were considered. Table 1.3 displays the risks examined in this study. See Haksöz, Kandemir, Camcı (2012).

We asked three basic questions:

- Which supply chain risks are more widespread in this time period (1999–2011)?
- Which risk has the most impact on global supply chain operations?
- Which industries are the most vulnerable? And to which supply chain risks?

During 1999–2011, 458 supply chain disruption incidents were identified using the LexisNexis database. The evolution of the frequencies of these risks is depicted in Figure 1.4. Among the three risks, the highest number of supply chain

Table 1.3 Global Supply Chain Risks Examined in the Study

• **Demand–Supply Mismatch:** Delays in product design, production, transportation, postponement, product shortage, stock-out, shortfall, unavailability, shipment delays, inventory write-off, raw material/component price volatility, commodity price volatility
• **Man-Made Disruptions:** product recall, breach of contract, contract termination, contract cancellation, strike, fraud, theft, sabotage, terrorism, espionage, product liability, labor disputes, port closure, political turmoil, geopolitical disruption, security breach, supplier default, bankruptcy, machine breakdowns
• **Natural Disasters:** Flood, fire, landslide, earthquake, hurricane, tornado, storm, pandemic, epidemic, extreme weather, climate change induced disaster

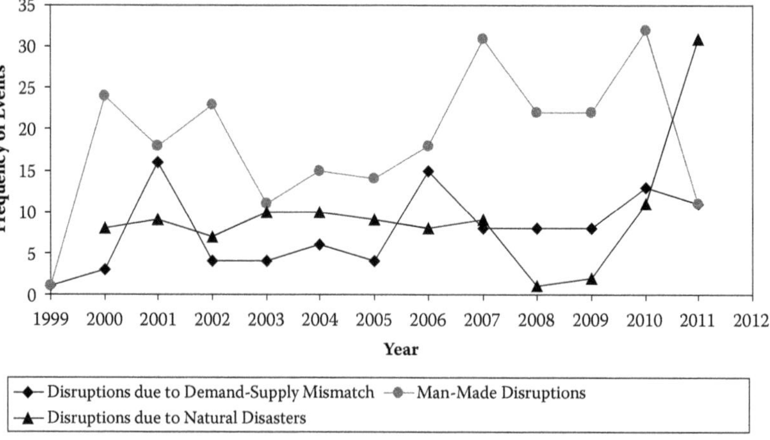

Figure 1.4 Frequencies of the global supply chain risk events during 1999–2011.

disruptions (242 of 458) was due to the man-made events. The next most affecting risk was natural disasters, with 115 events in the sample. Lastly, with 101 events, demand–supply mismatch was the least affecting risk among others. As discussed earlier, the number of natural disasters has increased in the past few years; hence, this result is not unexpected.

When one examines the particular risks that had the most impact, the results are more striking. In the demand–supply mismatch risk category, the top three risks having the highest frequency are out-of-stock (51.5%), delay (20.8%), and inventory write-off (19.8%).[29] The top three industries experiencing the highest negative impacts are pharmaceutical (23.8%), electronics (15.8%), and retail (8.9%). When we examine the second risk driver, that is, man-made disruptions, then the top three supply chain risks emerge as strike (30.1%), product recall (27.2%), and breach of contract (13.8%). In this class of events, the automotive (21%), food and beverage (13%), and information and telecommunication (8%) industries are the most affected. Finally, if we focus on natural disasters, the top three risks to global supply chain networks are fire (51.3%), snowstorm (19.1%), and flood (10.4%). In this category, the energy (26.7%), automotive (14%), and chemicals (9.4%) industries take the first three spots for the occurrence of these risks.

1.3 Recent Examples of Global Supply Chain Risks

After depicting the dark forest of global supply chain risks, now allow me to zoom in on a set of supply chain risks that adversely affected companies and industries in the past few years to provide you with a glimpse of the scale, scope, and consequences of such risks on global supply chain networks.

1.3.1 Out-of-Stock Risk

This risk is realized simply when a firm does not have a sufficient number of inventory on hand when the customers request them. If the customers are patient, they wait for back orders. If they are not, unless the product is something to wait for, potential sales are most probably lost. This case occurs when the product supply is insufficient for customer demand; thus, a supply–demand mismatch occurs. Out-of-stock risk occurred with the Apple iPhone in 2008. It has been reported that the iPhone's depleted inventory at Apple's U.S. and online stores, because of a component shortage, resulted in lost sales of up to 40,000 units a week[30] as demand. Similarly, Wii Nintendo game console's out-of-stock status during the U.S. holiday season (December 2007) was due to the inability of Nintendo manufacturers to meet the higher demand. This chain of events even created a market on eBay where price bids doubled. Patient customers were provided rain checks by the company with a promise of a January 2008 delivery.[31]

1.3.2 Delay Risk

Delay risk can occur in any process along the global supply chain networks. Sometimes delays can be contained within the system; other times, they may cause lost sales and even loss of market share. Surely other risks in the man-made and natural disaster categories might cause system delays. In the end, delay risk results in a demand–supply mismatch. In 2011, Boeing pushed back the first delivery of the 787 Dreamliner, already three years behind schedule, due to another delay caused by a fire in the electrical panel of the test aircraft. According to the news report, outsourcing most design and production of the parts for this aircraft created sometimes poor results and thus required rework by Boeing. In this process, all these delays cost penalties of billions of dollars to customers and suppliers.[32]

1.3.3 Strike Risk

Strikes are becoming widespread in many industries worldwide. Potential capacity reductions can be one culprit for striking. In the tire industry,[33] Goodyear experienced such a strike affecting sixteen U.S. plants in 2006. The strike was due to the firm's decision to reduce production capacity. Although closing plants would have saved the firm US$50 million, this strike did cost about US$2 million a day.[34] In the motorbike industry, Harley-Davidson has lived through a strike at its manufacturing and assembly plant in York, Pennsylvania. This strike decreased the quarterly shipment targets of the firm by 5,500 units. Hence, the company had to reduce parts (e.g., engines, transmissions) production at its other facilities. The revenue loss for this strike was expected to be US$11 million a day for Harley-Davidson.[35]

1.3.4 Product Recall Risk

Product recall risk creates trouble, particularly for global companies that manage a variety of suppliers worldwide. In 2007, the largest toy company Mattel had to recall over 18 million toys due to lead-tainted toys as well as design flaws in magnetic cars. While initiating such a recall decision is very difficult and most certainly costly, we have also seen that the Mattel chief executive officer Bob Eckert's swift reaction in calming customers and managing the crisis paid off well for the company in consecutive years.[36]

In our recent research on product recall risk management,[37] we demonstrate that the timing to take a product recall decision is a highly strategic one that needs to be thought through. In such a decision, the main trade-off is being too early or too late. If you are too early, you may make a recall decision that is unnecessary. The problems may have just occurred in a number of products, perhaps only a certain batch. However, if you are too late in making that decision, the problem might spread to many geographic regions and potential accidents could occur with consecutive legal court cases. Lateness could, in effect, kill a company in terms of reputation and customer goodwill. Such a late decision led Ford and Firestone to lose over US$1 billion for a tire recall in 2001. Furthermore, when two companies are involved in such a crisis, problems exacerbate. Both firms' stock prices were negatively affected due to this product recall.[38] So, what can firms do? A relatively brief yet concise answer I proposed in the *Wall Street Journal* is as follows:

> Firms can and should be more vigilant in reading and understanding the early signs of quality problems before crises arise. This can be done successfully only if there is a formal early detection system in place, which examines the critical leading indicators of quality problems both internally (with suppliers, partners and complementors) and externally (customers and the markets relevant to the products). But this is easier said than done. On the demand side, the firm has to have an effective communication framework with sales people in the field, dealers, regional distributors and global partners if the products are marketed globally. Early warning signals have to be correctly detected and acted upon to minimize potential value losses. On the procurement side, which has caused many of the recent recalls, suppliers (and their own suppliers) have to provide almost instantaneous updates on the quality levels of the products.[39]

Suppose you are the general manager of a global manufacturer. You have received some complaints regarding the malfunction of a certain product sold overseas. Your initial investigation reveals that the case turned out to be more serious than you initially thought. You start to contemplate the recall decision for the product, and a key set of questions pose themselves that need to be answered before taking action:

- What is the expected cost of product replacement in such a recall?
- What is the number of individual products to be replaced in such a recall?
- What is the spatial scale in this product recall (local, regional, global)?
- What are the likelihood and cost of potential litigations and court cases?
- Can we estimate the negative market reaction to the recall? Are we ready for such an undesirable reaction?
- Do we know the probability of manufacturing and service defects in our processes?
- Have we considered the involvement of suppliers in the product design and development? What are the chances that they could have caused undetected defects?
- What could be the potential impact (reputational risk) on our brand image of such a recall?
- What are the likelihood and severity of accidents involving these products?

1.3.5 Breach of Contract Risk

Breach of contract risk is defined as an operational risk under the *Clients, Products, and Business Practices* category of the Basel II framework.[40] Peabody Energy Corporation, one of world's largest private-sector coal companies, reported US$34 million losses due to a breach of contract breach by one of its suppliers in 2005.[41] Sometimes, intentional breach of contract may cause much larger costs to both parties due to long legal battles. For example, in 2001, Visteon Corporation paid American Axle & Manufacturing Inc. an arbitration award of nearly US$14.9 million for violating an agreement to buy forgings.[42]

For firms that produce or purchase commodity products such as metals, wheat, cotton, natural rubber, and paper and pulp and use them in their core operations, the spot market price plays an important role in the realization of the breach of contract risk. The intelligent use of increasingly transparent and liquid spot markets worldwide is quickly becoming a strategic priority for such firms.[43] The movement of the spot market price affects how buyers and sellers of the commodity will behave with respect to their supply contract commitments.

A few examples from the steel supply chain illustrate my argument. The largest steelmaker in the world based on production capacity, ArcelorMittal, stated in 2008 that only 20% of its capacity is allocated to long-term fixed price contracts. The larger chunk is devoted to spot market selling where there was a favorable price.[44] On the other hand, moving upstream in the steel supply chain, the iron ore buyers turned to the iron ore spot market that offered lower prices than the contracts. For example, in November 2008, the Australian iron-ore producer Mount Gibson Iron Ltd. stated that three of its iron ore customers had breached their contracts to buy the commodity from the spot market.[45] Surely, an intentional breach may entail penalties for the breaching party. Breach penalties must be carefully designed considering potential legal consequences.[46]

To mitigate such potential breaches of contract, BHP Billiton, a South American mining company, has started to sell iron ore on quarterly contracts with a price linked to the spot market.[47] In our research,[48] we show that breach of contract risk decreases in the presence of a price renegotiation option. A price renegotiation option can be formally designed at the beginning of the supply chain contract—as a product, as opposed to going through an uncertain renegotiation process—that allows the buyer–seller to revise the supply chain contract price at a fixed date based on the respective spot price. We demonstrate that designing a supply contract with a *bundled option*, which is composed of not only *price renegotiation option* but also *contract abandonment option*, creates more value for the contract than either of the options in isolation. In highly volatile spot markets, supply contract value increases with such a bundled option. In the presence of high spot price volatility, the bundled option is more valuable when the renegotiation date is selected to be closer to the half-life of the contract. Thus, BHP Billiton's aforementioned intelligent contract pricing strategy could be formalized for other commodity firms in different markets using an effective framework, providing multiple flexible options such as price renegotiation and contract abandonment in the contract. This type of flexibility, as long as it is priced correctly, has been shown to enhance the contract value for both partners and mitigate the breach of contract risk.

1.3.6 Weather Risks

Weather risks can be classified into *catastrophic* and *noncatastrophic*.[49] A catastrophic weather risk includes the damages by flood (Thailand floods in 2011–2012), tsunami (Japanese tsunami in 2011), and hurricane (Hurricane Katrina in 2005). A noncatastrophic weather risk may be extreme heat or cold, a snowstorm, big rain or snowfall, frost, or drought. Even though catastrophic weather risks make the headlines, such as the recent floods in Thailand and Australia, the consequences of the noncatastrophic weather risks could be enormous for different industries. A colder or warmer winter can affect the profitability of energy, apparel, and ski resort firms. A cooler summer may reduce the revenues for a beverage producer.[50] In fact, 160 of 200 companies defined weather as a key aspect in their operations.[51]

In winter 2010, the northeastern region of the United States experienced a number of snowstorms that disrupted the flow of goods in different supply chains. For example, the *Platts Coal Outlook* reported on the adverse impact of snowstorms on companies such as CSX and Norfolk Southern's railroad deliveries to coal-fired power plants:

> Both CSX and Norfolk Southern reported delays of 72 hours or more during points of the week ending February 7 after back-to-back storms dropped up to several feet of snow on parts of Maryland and

> Pennsylvania. Many other areas received more than a foot as well as tropical storm-force winds that created whiteout conditions. Norfolk Southern coal shipments plummeted 28% during the week of the storms, adding 3.5 percentage points to the year-to-date losses the railroad reported February 18 [...] CSX experienced a similar, though less pronounced, jolt to its utility coal shipments, which fell 23% during the week....[52]

A number of snowstorms in İstanbul dramatically affected land, sea, and air transportation in February 2012. In one such snowstorm, 183 flights (domestic and international) were canceled by Turkish Airlines.[53] Sea and public transportation was also disrupted due to poor visibility. In addition, disruptions in electricity distribution were reported. A public–private consumer collaboration is a must to manage such weather risks. Great lessons can be learned from good practices such as those demonstrated in New York City and Chicago, where snowstorm management is already in the genes of public authorities. The City of Chicago, for example, has developed a real-time snowplow tracker[54] to share this critical information with its inhabitants.[55]

1.3.6.1 How the Price of Cherries Are Affected by Honeybees

In June 2011, it was reported that honeybees in Turkey could not pollinate cherry flowers due to unexpected springtime weather conditions.[56] Spring is the traditional blooming period for cherry trees. It is interesting to note that if there is more rain, humidity, and cooler weather, the honeybees cannot work effectively. The pollination needs to occur within 24 hours of blooming. Yet this time period did not exist for the honeybees in 2011. Hence, it was expected that cherry production would be halved and prices would increase. Business executives also need to be aware of the *weather–fauna–flora* interactions in food supply chain networks: not only knowledge of weather conditions, but also the behavior of other animal and plant members that share this planet with us. Cascading risks and interdependencies need to be understood precisely and required steps taken wisely to manage such risks in food supply chain networks.

1.3.6.2 Limited Supply of Red Roses in Turkey for Valentine's Day

A Siberian cold front and subsequent snow damaged red rose production in Turkey, reducing it by 60% in 2012. According to news reports, Valentine's Day lovers experienced difficulty in finding the red roses they wanted to present to their loved ones.[57] These events forced companies to import roses from Ecuador and the Netherlands. Although red rose prices shot up 50%, supply disruptions led to 10% lower expected revenues than those of the previous year.

1.3.6.3 Swimming in the Flood, Drowning in the Losses

Flood risk and its catastrophic impact have been recently witnessed by the global community in the landfall of Tropical Storm Nock-ten, which created huge floods for 175 days in Thailand.[58] By March 2011, Thailand had experienced 344% more rainfall than in an average year. The World Bank estimated costs of up to US$45 billion, 815 deaths, and 13.6 million people impacted. [59] Global firms such as Apple, Broadcom, Cisco, Canon, Dell, Ford, Hewlett-Packard, Hitachi, Honda, Intel, Lenovo Group, Panasonic, Toyota, Western Digital, and many others were adversely affected by these floods.[60] More than 10,000 factories were closed, and more than 350,000 workers were laid off due to production disruptions.[61]

1.4 Turkey and Supply Chain Risks: From History to the Future

> A timid merchant neither loses nor makes a profit.
>
> **—Turkish Proverb**

Let me provide a few highlights regarding the Turkish economy before discussing supply chain risks. Turkey is the sixteenth largest economy in the world and one of the fastest growing countries with a compound annual growth at constant prices of 4.8% from 2002 to 2010. Turkey is classified as an emerging economy by the International Monetary Fund (IMF) and the World Bank, whereas the *World Factbook* of the Central Intelligence Agency (CIA) identifies Turkey as a developed country.[62] In financial circles, Turkey is addressed in the special group of countries called the BRIC-MIST, a group composed of Brazil, Russia, India, China, Mexico, Indonesia, South Korea, and Turkey. [63]

The sector composition of the Turkish economy is dominated by services (64%), followed by production (27%) and agriculture (9%.) The majority of supply chains in Turkey include manufacturing, transportation, storage, communication, and trade.[64] Interestingly, İstanbul comes fourth after New York, Moscow, and London as the city that the highest numbers of billionaires identify as their place of home around the globe.[65]

In terms of cultural behavior and attitudes, risk management has appeared as part and parcel of Turkish society and its DNA for centuries. If one looks at the nation historically through the centuries of its existence, one could characterize the nomadic lifestyle of Turkish people as creating the response for intelligent risk management approaches. In such an itinerant, transitory existence, one needs to be adaptive, responsive, and resilient in various weather, geographical, and government conditions. Individuals need to frugally manage their resources for their family, domestic livestock such as camels and horses (used for transportation), and sheep and goats (used for food and products to sell).[66] Many Turkish tribal migrations occurred more than a thousand years ago from the Central Asian steppes toward the west, north, and south during extended time periods. In particular, Turkmen tribes[67] migrated toward today's Anatolia in the face of the Mongol empire moving west and crushing cities along its path.[68] During such arduous, dangerous, and challenging migrations, it is obvious that risk sensing and perception as well as risk taking would have been honed. In short, I firmly believe that Turkish citizens carry a risk compass due to their DNA and upbringing. Whether they practice it well is another question, of course.

In the Byzantium, Turkish Seljuki, and Ottoman Turkish states, the major parts of the historic Silk Road wound through Anatolia (Asia Minor). Beginning with the Seljuki Turkish state, caravanserais (*kervansaray*) (see Figure 1.5 for one such caravanserai standing for eight centuries in Turkey) were built along the Silk Road to smooth the flow of goods, people, information, and ideas. These lodgings accommodated caravan travelers with their camels (as well as other animals such as horses and mules) and goods for a few nights. In the open countryside, these caravanserais also played the role of markets.[69] One of the major risks along the Silk

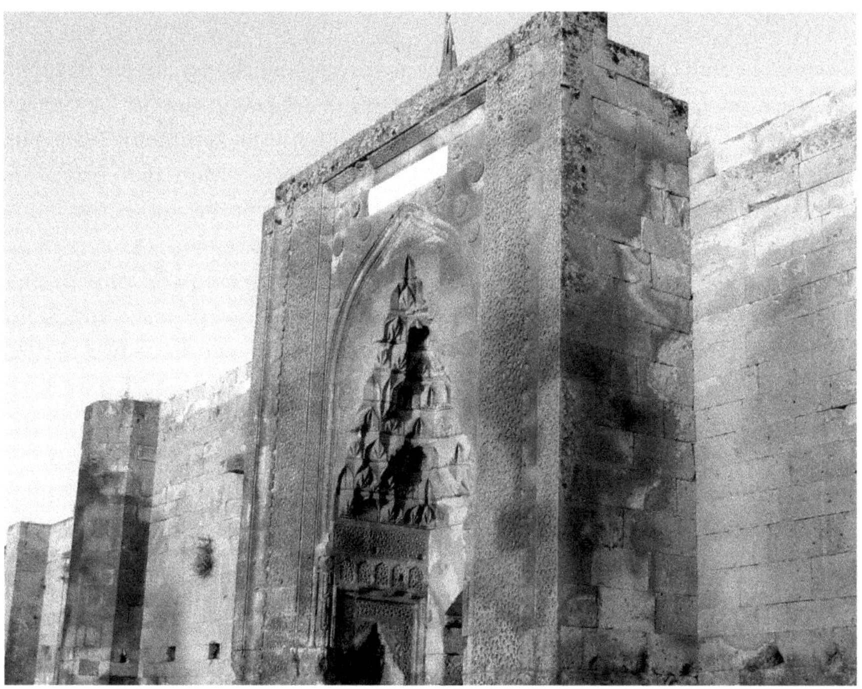

Figure 1.5 Ağzıkarahan caravanserai and its main gate with unique inscriptions and motifs on the Silk Road, Aksaray, Turkey (thirteenth century, Turkish Seljuki period). (Photo by the author.)

Road supply chains was thus caravan robberies. The Ottoman historian Suraiya Faroqhi provides an interesting account of this risk along the Silk Road:

> For nomads in outlying areas, robberies might be an emergency source of income when an epidemic had carried off part of their sheep or a drought had destroyed the harvest which they expected from fields in their winter quarters. In other cases, the threat of robbery might be used systematically to extract tribute payments from caravans passing through the territory controlled by the tribal units.[70]

1.5 Supply Chain Risk Perception of Turkish Professionals

> When playing Russian roulette, the fact that the first shot got off safely is little comfort for the next.
>
> **—Richard P. Feynman**[71]

Recently, I conducted exploratory research to understand the perception of supply chain risk by Turkish professionals.[72] The study examined a number of strategic industries in Turkey, which included automotive, high-tech, commodities (metals/paper), fast-moving consumer goods, and pharmaceutical.[73] More than half of the supply chain professionals (54.25%) surveyed emerged from the automotive industry. Figure 1.6 illustrates the composition of the rest of the sample. The participant pool was also diverse in terms of job titles, with supply chain and procurement directors and managers having more than two decades of experience and junior supply chain, procurement, and logistics specialists having as little as two to three years of experience.

One major goal of this study was to identify the risk events that severely affected the supply chain professionals' perceptions. These risk events were classified into three categories: (1) supplier failures (e.g., their costs, frequency, the need to find alternatives); (2) late deliveries; and (3) quality problems in the shipments. The second goal was to identify the risk hedging strategies actively used. Figure 1.7 displays the major risk hedging strategies—both operational and financial—used in practice. Overall, the most popular risk hedging strategies were multiple sourcing, formulating business continuity plans, purchasing fixed-price supply contracts, designing penalties for nonperformance and breach of contract, and reserving extra supplier capacity. Surely, different industries have a predilection for a certain risk hedging strategy. Multiple sourcing seems to especially be the most popular risk hedging strategy. To better understand the significance of multiple sourcing strat-

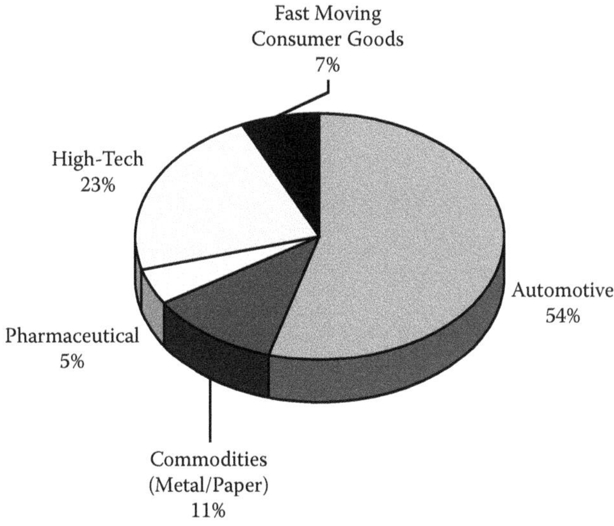

Figure 1.6 Composition of the industries surveyed in the supply chain risk perception study.

egy as well as its counterpart strategy (i.e., single sourcing), let me provide a few examples from practice.

1.5.1 Multiple versus Single Sourcing Strategy

The sourcing strategy from multiple suppliers with different risk profiles has been used by many industries and firms. In fact, Part 3 will show that this supply chain risk hedging strategy is currently employed with certain nuances by all of our leading companies, Kordsa Global, Brisa, and AnadoluJet (a brand of Turkish Airlines).

Using this risk hedging strategy, firms can mitigate the supplier failures and unexpected supply chain disruptions using multiple suppliers, preferably located in different regions and locations. Once this strategy is employed, total supply capacity increases, yet the cost of maintaining multiple suppliers simultaneously goes up. Reserving the appropriate amount of capacity in different suppliers and allocating the supply volume thus become key problems to be resolved. Some firms would prefer to use a single dedicated supplier due to several reasons such as cost efficiency (due to economies of scale of investment), quality, and delivery timing risks.

There is a clear trade-off between cost efficiency and responsiveness while deciding on a single versus multiple sourcing strategy. The Aisin fire that occurred in one of the key suppliers of Toyota[74] in 1997 taught many lessons on how supplier failure risks should be managed. Tsuyoshi Mochimaru, an auto analyst at Nikko Research Center, argued that the priority of Toyota shifted toward cost reduction instead of risk mitigation, especially after 1990, as the domestic market became

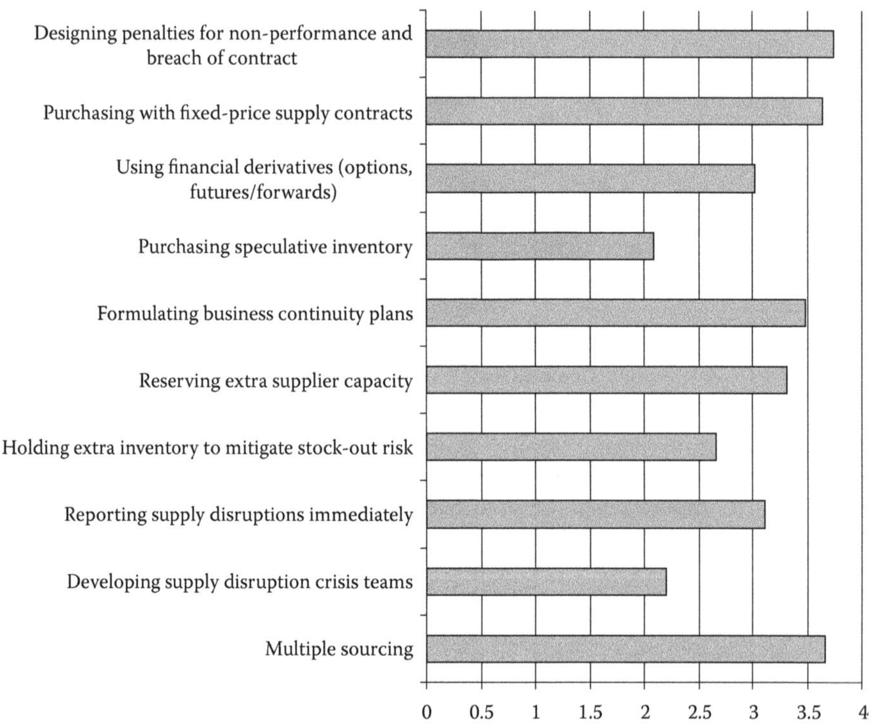

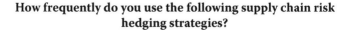

Figure 1.7 Frequency of use for supply chain risk hedging (operational and financial) strategies (mean values are given using a Likert scale, 1–5, where 1 = never, 2 = seldom, 3 = sometimes, 4 = somewhat frequently, and 5 = frequently.)

more competitive and the yen appreciated. Hence, the cost reduction objective led to single sourcing, which had a negative effect in the Aisin fire incident. [75]

In 2002, Land Rover had an issue with insolvency by some of its suppliers. Steve Jones, the director of purchasing for Land Rover, said that a single supplier strategy was preferred due to investment and quality concerns. According to Jones, working with a single supplier incurs lower investment cost and also leads to lower variability in product quality. [76]

On the other hand, using a single dedicated supplier could be valuable in terms of creating a long-term intimate relationship that may enable productive new product development and the sharing of market intelligence among others. Yet this loyalty also brings with it potential costs as switching to another supplier grows much more costly due to a *lock-in process*. Lock-in situations in buyer–supplier relationships have been studied using social exchange theory.[77] If a buyer is locked into a supplier relationship, he or she will be exposed to supplier failure risks more than will a buyer with loose relationships with suppliers. So a lock-in process increases both switching costs and risk exposure of the buyer.[78]

Not only the lock-in process but also other more tangible reasons exist for the preference of dual or multiple sourcing strategies over single sourcing. One of the most immediate is the increasing frequency and impact of extreme natural disasters. For example, in the recent Japanese Fukushima disaster, aerospace companies such as Boeing began considering to dual source some of the subassemblies and components (such as fasteners) of the aircrafts. This step appears to be appropriate, considering that the Japanese suppliers comprise 35% of the 787 and 20% of the 777 Boeing models.[79]

Apart from natural disasters, industrial accidents in suppliers may also result in supply chain disruptions unless secondary sources are available. A local firm in Ireland, Smurfit Kappa Ireland, experienced a burst pipe accident in its Belfast plant. Having a second supplier available in Lurgan enabled the company to immediately switch production and prevent the potential impact of this accident onto its customers.[80]

One other reason for multiple suppliers is to mitigate strike risks. In 2009, Boeing decided to dual source the 787's vertical stabilizer and other assemblies to mitigate the risk of strikes.[81] One year before this decision, the International Association of Machinists and Aerospace Workers (IAMAW) Local 751 experienced a two-month walk-out that disrupted the Boeing production line. The investment cost to start a second facility in Charleston, South Carolina was estimated at US$750 million to US$1.5 billion. On the other hand, if a strike hit the company, with supplemental payments, the cost could easily spiral to US$5 billion.[82]

Before closing this chapter, let me provide the strategic insights derived from this empirical study:[83]

- There is a need to integrate the operational and financial hedging strategies in supply chain risk management.
- Firms procuring commodities with high price volatility should
  - Use financial derivatives, but with great care.
  - Invest more time and energy in designing flexible supply chain contracts with intelligent real options such as abandonment and renegotiation to mitigate breach of contract risk.
- Supplier collaboration is essential in increasing supply chain resilience, especially in crisis situations. Hence, win–win solutions must be sought in every encounter with suppliers.[84]
- Firms need to develop market intelligence teams that should mine different data terrains such as
  - Historical supplier performances
  - Evolution and dynamics of supply and customer markets
  - Trends in supply chain contracts

- Constantly work on enhancing the supplier diversity in terms of
  - Production volumes and timings
  - Size and capability of firms
  - Onshore, offshore, and near-shore considerations
- Be aware of the potential downside risk of supplier lock-ins as relationships strengthen:
  - Always be ready to walk away from a relationship that is damaging to both parties.
  - Plan on developing a dual or multiple sourcing strategy in case the locked-in supplier is unique.
- Never use a single sourcing strategy just to enhance the quality and on-time delivery performance:
  - Quality and delivery timing risks need to be managed in conjunction with your single, dual, or multiple sourcing strategy.

After examining a variety of supply chain risk examples from different regions and industries around the globe, I summarize major supply chain risks, their impacts on global supply chain networks, as well as key hedging and mitigation strategies in Table 1.4.[85] This summary is presented considering the process view of the supply (i.e., supply–process–demand) chain as well as the external environment. Figure 1.8 displays the schema of such a perspective and also shows the risk interdependencies that will be discussed in detail in Part 2.[86]

Part 3 discusses how three risk intelligent supply chains in Turkey thrive in this fragile, capricious, and entangled world. But, before we learn lessons from leading Turkish companies, let us first build the fundamental underpinnings of risk intelligent supply chains in Part 2. This is the next stop in our journey.

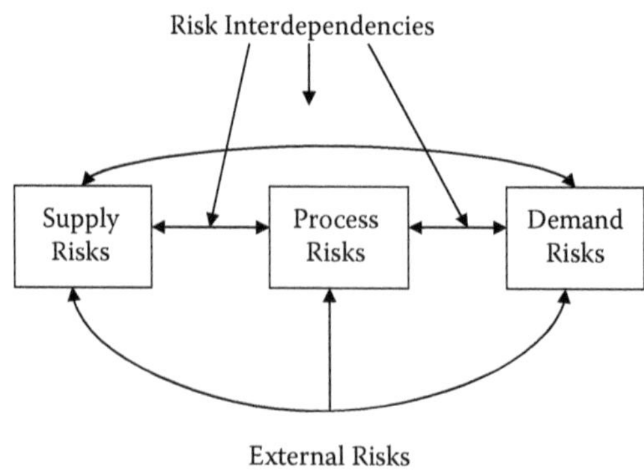

Figure 1.8 Process view of supply chain network risks.

Table 1.4 Major Risks and Mitigation Strategies in Global Supply Chain Networks

Risk Types	*Major Impact on Global Supply Chain Networks*	*Mitigation Strategies*
Supply Risks	Delays and congestion in supply pipeline	Use of backup and safety inventory, capacity, and suppliers[88]
	Minor and major accidents in suppliers and logistics providers	Fail-safe design Taguchi principles, near misses, precursors, early warning systems
	Breach of contract and nonperformance of suppliers[89]	Design of smart and flexible options written in contracts[90]
	Insolvency or bankruptcy of suppliers (e.g., tier 1, 2, 3)	Insurance for insolvency, trade credit financing of suppliers
	Inferior quality items	End-to-end robust quality management system, supplier auditing
	Increase of supply prices as well as price volatility	Portfolio of contracts such as long-term and option contracts, as well as use of spot markets, purchase of raw material sources (e.g., African mines[91])
Process Risks	Internal machine, material, human-related failures, breakdowns	Better preventive maintenance and system design
	Internal plant accidents	Use of near misses, precursors, and early warning systems
	Loss of critical talent to competitors	Continuous monitoring and development of key talent (better talent management)
Demand Risks	Demand–supply mismatch	Operational hedging tools such as postponement and delayed differentiation, capacity reservation

(continued)

Table 1.4 Major Risks and Mitigation Strategies in Global Supply Chain Networks (continued)

Risk Types	*Major Impact on Global Supply Chain Networks*	*Mitigation Strategies*
Demand Risks (cont.)	Abrupt change of competition with innovative products and services	Peripheral monitoring of competition and markets[92]
	New entry to the market	Effective of use experimentation[93] and quick time-to-market, market testing with lead users
External Risks	Natural disasters (earthquakes, hurricanes, floods, thunderstorms)	Weather derivatives,[94] purchase of insurance, catastrophe bonds, catastrophe-linked options[95]
	Fires at supply, process, demand processes	Insurance as well as operational hedges such as sprinklers
	In offshoring processes (loss of intangibles and intellectual property fraud)	Effective management of lock-in and switching processes
	Social risks (labor disputes, strikes, child labor), political and regulatory risks (European Union regulation, wars, political unrest)	Smart supplier auditing,[96] market intelligence for political events

2

HOW TO BECOME A RISK INTELLIGENT SUPPLY CHAIN

Chapter 2

Risk Intelligent Supply Chains and the I-Quartet Model

When there is an unknowable, there is a promise.

—Thornton Wilder[1]

To achieve anything, you must be prepared to dabble on the boundary of disaster.

—Stirling Moss[2]

Risk is no longer something to be faced; risk has become a set of opportunities open to choice.

—Peter L. Bernstein[3]

2.1 Introduction to Risk Intelligent Supply Chains

In Chapter 1, I discussed the status of the global supply chain networks and their fragility. I also presented the supply chain risk perceptions of business executives in Turkey. It should now be clear that a diverse set of risks from all directions has been continuously hurting supply chain networks and the companies that manage them. Business executives point out that they need fresh perspectives to be able to withstand any type of potential calamity in the future that threatens global supply chain networks.[4]

After years of research, consulting, and teaching in North America, the United Kingdom, and Turkey regarding supply chain risk management, my response to this challenge is to develop the concept of *Risk Intelligent Supply Chain* (RISC). A like mind, Edward de Bono, states, "Concepts are important in generating ideas and designing ways forward. Where there is no routine available, concepts are essential."[5] To make RISC operational, I created a conceptual model coined the *I-Quartet model.* The metaphor behind this concept is a *string quartet,* where harmonious beautiful compositions are performed by four string players—usually two violinists, one violist, and one cellist. Likewise, the I-Quartet model roles work in harmony to achieve a risk intelligent supply chain. This part of the book—Chapters 3 through 6—will provide details on this conceptual model.

Before an overview of the I-Quartet model is offered, it is useful to begin with a number of definitions on *intelligence* and *risk intelligence.* To explain this phrase, first we need to examine intelligence. The root of intelligence, *intelligo,* means to "select among."[6] You need a set of ideas, tools, methods, and concepts from which you can select. In other words, you make novel and beneficial combinations among disparate sets of ideas, concepts, and thoughts. When one looks at the theory of intelligence, it is imperative to mention the theory of multiple intelligences, which was developed in 1983 by Howard Gardner. According to Gardner, each individual carries a set of cognitive abilities, not just one ability: spatial, linguistic, logical-mathematical, bodily-kinesthetic, musical, interpersonal, intrapersonal, and naturalistic.[7] Alongside other researchers in the field of cognitive intelligence, Tony Buzan, the inventor of Mind Maps, offers in his book *Head First* ten types of intelligence: creative, personal, social, spiritual, physical, sensual, sexual, numerical, spatial, and verbal intelligence.[8] Augmenting these lists, in 2004 Raimo Hämäläinen and Esa Saarinen introduced and defined the concept of *systems intelligence* as follows:

> ... Intelligent behavior in the context of complex systems involving interaction and feedback. A subject acting with systems intelligence engages successfully and productively with the holistic feedback mechanisms of her environment. She perceives herself as part of a whole, the influence of the whole upon herself as well as her own influence upon the whole. By observing her own interdependence in the feedback intensive environment, she is able to act intelligently.[9]

Surely, intelligence as a cognitive ability or potential is necessary but not sufficient in practice. Intelligence needs to be honed and used effectively with a thinking skill. Edward de Bono beautifully expresses the difference between intelligence and thinking skill:

> I like to think of intelligence as being equivalent to the horsepower of a car. The skill of thinking is then the skill of the driver. There may be a powerful car driven badly and a humble car driven well. Indeed, a powerful car may be particularly dangerous because it demands a higher degree of driving skill. I believe that thinking skill, like driving skill, can be developed through training and deliberate effort.[10]

Furthermore, even if the thinking skill is available, development, according to Russell Ackoff, may not occur unless wisdom operates in the process. In differentiating between intelligence and wisdom, Ackoff puts forward a very important claim:

> The efficiency of behavior or an act is measured relative to an objective by determining either the amount of resources required to obtain that objective with a specified probability, or the probability of obtaining that objective with a specified amount of resources. The value of the objective(s) pursued is not relevant in determining efficiency, but it is relevant in determining effectiveness. *Effectiveness is evaluated efficiency.* It is efficiency multiplied by value, efficiency for a valued outcome. Intelligence is the ability to increase efficiency; wisdom is the ability to increase effectiveness. The difference between efficiency and effectiveness—that which differentiates wisdom from understanding, knowledge, information, and data—is reflected in the difference between development and growth. Growth does not require an increase in value; development does. Therefore, development requires an increase in wisdom as well as understanding, knowledge, and information.[11]

Intelligence, thus, cannot be studied independent of reference to wisdom. I will therefore discuss the value of wisdom at the beginning of the Part 3 before presenting the cases of leading Turkish companies.

2.2 Risk Intelligence versus Risk Intelligent Supply Chain

> It takes something more than intelligence to act intelligently.
>
> **—Fyodor M. Dostoyevsky[12]**

The term *risk intelligence* is not my own creation. As a phrase it has consistently appeared in three books. David Apgar (2006), in *Risk Intelligence: How to Live with Uncertainty,* defines the risk IQ and ways to improve this intelligence by suggesting simple rules, that is, recognizing and identifying the correct risks, pursuing projects one at a time based on their riskiness, and building effective networks of partners, suppliers, and customers to manage a diverse set of risks. In *Surviving and Thriving in Uncertainty: Creating the Risk Intelligent Enterprise,* Rick Funston and Steve Wagner (2010) studied the calculated risk taking as the main topic of interest. To that end, Funston and Wagner suggest ways to intelligently manage enterprise risks, some of which are challenging business assumptions, defining risk appetite, and anticipating causes of failure to enhance preparedness. Last, Dylan Evans (2012), in his book *Risk Intelligence: Learning to Manage What We Do Not Know,* sheds light on this special type of intelligence that deals with uncertainty and doubt.

To date, neither any of these mentioned books nor any scholarly work has focused on *supply chain risk intelligence* per se. Thus, this is the first book introducing the concept of risk intelligent supply chain. Supply chain risk intelligence is not a set of skills or cognitive abilities. Supply chain risk intelligence is a mindset composed of four interdependent and interacting roles. These four roles are formulated in the I-Quartet model as the *Integrator, Inquirer, Improviser,* and *Ingenious,* which will be discussed next. Second, this book presents the specifics of successful risk intelligent supply chain cases from Turkey that have the potential to inspire firms worldwide. Surely, around the globe, there are many unrecognized champions in supply chain risk intelligence. This book is the first humble step in a challenging risk intelligent supply chain journey. The journey will continue with my ongoing work in the next few years; I hope and anticipate that this research will spur other studies in this novel context.

Before I provide an overview of the I-Quartet model, there are at least three fundamental questions every business executive has to answer while contemplating a risk intelligent supply chain:

- Can we design and operate a risk intelligent supply chain that is immune and also adaptable to various types of risks?
- What are the critical internal and external factors as well as dynamics we need to consider in this process?
- What managerial levers can we use to continuously increase the risk intelligence of our supply chain network?

2.3 The I-Quartet Model: A Synopsis

In this section, I provide a synopsis of the I-Quartet model. In short, it explains the various roles that a risk intelligent supply chain plays under different conditions. These four necessary roles—Integrator, Inquirer, Improviser, and Ingenious—have to be played one way or another by any risk intelligent supply chain. These four roles, like four string instruments in a quartet, cannot act independently. They are highly interdependent and interactive. They also need to create harmony. At a particular time, two specific roles may take the lead, whereas at other times three different roles must simultaneously act. For example, in Chapter 8 we will see that Kordsa Global plays the roles of Integrator, Improviser, and Ingenious while managing its supply chain contracts. Brisa, discussed in Chapter 9, plays the roles of the Inquirer and Improviser while managing political risks in its supply chain network. These interactions and interdependencies also create complexity that cannot be overlooked.

Figure 2.1 displays the I-Quartet model and its four roles as well as the interactions between these roles. In the model, the four roles interact. These interactions will be elaborated in upcoming chapters:

1. **Integrator–Inquirer**: You learn through the Inquirer role and revise the strategies in the Integrator role while managing the supply chain network.
2. **Inquirer–Improviser**: As you learn in the Inquirer role, your supply chain network becomes more resilient. Likewise, as the Improviser role is improved by enhancing its characteristics, the Inquirer role becomes more valuable. You manage the *Swan Lake* risks much more effectively.
3. **Improviser–Ingenious**: As you become a better Improviser, your risk-taking ingenuity increases. For example, as the diversity increases in the supply chain network, your probability of getting a positive black swan increases. Yet after a certain threshold, the complexity of the supply chain network may result in more negative black swans as will be explained later in Chapter 6. You become more comfortable with *fork-tailed risk taking*, considering both types of black swans intelligently.
4. **Ingenious–Integrator**: As you become more Ingenious in risk taking, you can better orchestrate the supply chain network in the Integrator role. To this end, you know better the weakest spots to focus on and improve the total value of the supply chain network.
5. **Integrator–Improviser**: These roles interact such that the creativity, diversity, flexibility, and tinkering features of the supply chain network are being constantly revised to enhance the Improviser and Integrator roles, an interactive process that gradually increases the supply chain network resilience.
6. **Inquirer–Ingenious**: These roles interact such that ingenuity in risk taking is continuously rejuvenated by learning from failures, near misses, precursors, and weak signals.

In short, to become a risk intelligent supply chain, one should orchestrate the global supply chain network while discovering, exploring, learning, and adapting constantly. In this process, one also needs to be a great improviser to achieve resilience as well as ingenious in risk taking.

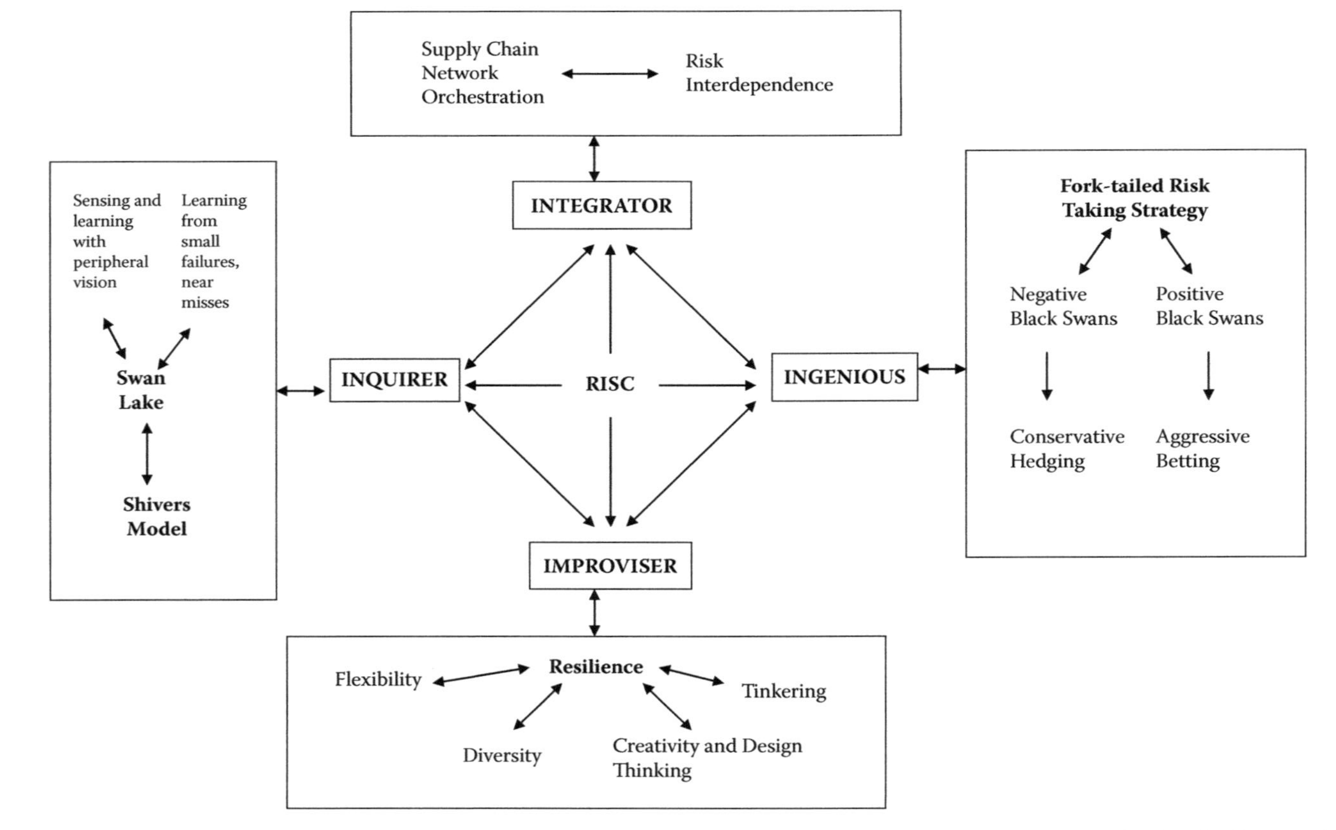

Figure 2.1 The I-Quartet model (roles and interactions).

2.4 The Acrobat over the Bosphorus

Before we continue to discuss the details of these roles one by one, I want to wrap up this chapter with a metaphor, *the Acrobat over the Bosphorus*. The book's cover art introduces this acrobat, and a number of drawings at the beginning of each I-Quartet model role chapter are created based on the same metaphor. All of these drawings include an acrobat on a single tightrope or a network of tightropes. This metaphor, the Acrobat over the Bosphorus, has been developed for this book to symbolize the risk intelligent supply chain concept within its particular geographic context and its place of origin.

Sensing, perceiving, assessing, mitigating, hedging, and managing supply chain risks in Turkey are very much similar in spirit to the acrobat who crosses the Bosphorus in İstanbul on a tightrope under any weather and sea traffic conditions, which might be pretty nasty at times.[13] This is why the original drawing of this book shows such an acrobat overlooking İstanbul, the crossroads of Asia and Europe, while trying to safely traverse the tightrope.

In the I-Quartet model chapters (Chapters 3–6), this metaphor has been used to create various different acrobats who play the particular I-Quartet role of the relevant chapter. I would like to leave to the imagination of the readers why these acrobats traverse the tightrope, that is, act the way they do in these drawings. Table 2.1 summarizes the strategic actions taken by an acrobat in the four I-Quartet roles.

Table 2.1 I-Quartet Model of an Acrobat over the Bosphorus

The I-Quartet Model Roles	*Strategic Actions of an Acrobat*
Integrator	• Orchestrate all of your skills, abilities, and experience in the context of external–internal conditions to smooth your movements
Inquirer	• Sense, perceive, and learn constantly and immediately from your small failures and near misses • Anticipate continuously the potential problems (tightrope, weather, yourself) ahead
Improviser	• Increase resilience by creatively tinkering and improvising based on what is on hand • Be diverse and flexible in your movements
Ingenious	• Take risks intelligently while being open to joyful surprises

Chapter 3

The Integrator: Orchestrating an Interdependent Supply Chain Network

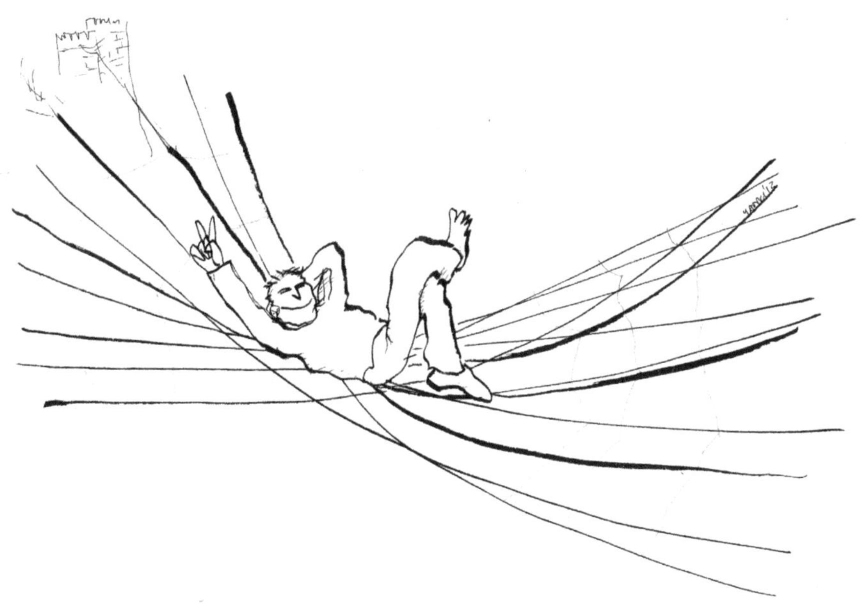

Sometimes people go ahead and divide an elephant in half anyway. You don't have two small elephants then; you have a mess. By a mess, I mean a complicated problem where there is no leverage to be found because the leverage lies in interactions that cannot be seen from looking only at the piece you are holding.

—Peter M. Senge[1]

A view of network against network means that the companies with access to the best networks not only can outperform competitors today, but also have the capacity to flexibly outperform them tomorrow. This is like the bench strength of an athletic team.

—Victor Fung, William Fung, Yoram (Jerry) Wind[2]

THE INTEGRATOR IN GLOBAL ECOSYSTEMS

Spiders are extraordinary integrators. They construct integrated webs, networks of their own, for their livelihood. Any small glitch in network integrity threatens spiders' lives. They use the webs to sense their environments and capture their preys.

THE INTEGRATOR ROLE IN BRIEF

The *Integrator* role integrates and orchestrates global and local supply chain networks with a multitude of stakeholders such as suppliers, customers, employees, and partners. It capitalizes on the strategic and super-additive values of networks while constantly reducing the flow frictions in material, cash, people, and information. Risk interdependencies in the supply chain network are managed intelligently, which, in turn, increases the robustness of the network.

3.1 Networks and Complexity

A network is a possibility factory.

—Kevin Kelly[3]

First, I briefly introduce the basic concepts for networks and complexity that will form the backbone of the analysis of the *I-Quartet model* and the risk intelligent supply chain.

Networks are composed of entities that are called nodes, clusters, and hubs connected by links.[4] Such nodes can be firms in supply chain networks, or individuals in social networks. There are different network topologies that are observable in practice.

The first network type is the *hub-and-spoke network*. In such a network, hub (central) nodes become intermediaries between other nodes. Because hub nodes control network interactions, network effects (in other words, complex behavior to be explained later in the chapter) are rarely seen. The second type is called the *random network*, where each node has an equal probability of connection to any other node.[5] Early research in networks had suggested that large-scale telephone networks were random. The *scale-free network* represents the third type; it can be found in nature (e.g., ecological networks) as well as on the Internet. The scale-free network is characterized by the fact that a few nodes have a large number of links and that most of the nodes have few numbers of links. In essence, these networks are shown to be *fractal;* that is, they are *self-similar* over scale.[6] And they also exhibit power laws meaning that some nodes are more important than others.[7] We will discuss power laws, in the context of the *Inquirer* role in the following chapter. Another important feature of scale-free networks is their paradoxical nature: robust under random attacks (natural disasters) and node removals, yet fragile under intentional attacks (sabotage and terrorist attacks) on the hubs (highly connected nodes). Accordingly, the preservation of the crucial nodes must be the main priority in scale-free networks to ensure robustness. Bearing this aim in mind is critical in designing and managing robust supply chain networks. Having identified the characteristics and weaknesses of the aforementioned network types, two real-world questions arise: Can stakeholders unite to design a more robust supply chain network, and how could this be done?

Looking at other networks might be a good starting point to embark upon these questions. For instance, the concept of the *strength of weak ties*, which has been developed in the context of social networks, states that *weak ties* more effectively bridge different networks than do strong ties. The presence of weak ties produces sparse (less dense) networks.[8] These networks have been proven superior in idea generation, learning, and creativity and can also reach a broader and diverse audience. As the networks become denser, mutual trust and support among individuals increase, a condition that eventually enhances the effectiveness of idea implementation and execution. Thus, there is an interesting balance between sparse and dense social networks that needs to be effectively managed.[9]

When we speak of networks, we need to understand another key concept—complex systems—as they are intimately connected. Scott Page provides a clear definition of a *complex system*: "[It] consists of diverse entities that interact in a network or contact structure—a geographic space, a computer network, or a market. These entities' actions are interdependent—what one protein, ant, person, or nation does materially affects others."[10] There is a difference between *complex* and *complex adaptive systems*. In complex systems, entities follow fixed rules, whereas in complex adaptive systems the respective entities can adapt to changing environmental conditions. The interactions of entities in such systems are nonlinear, dynamic, and probabilistic.[11] As entities in the network adapt, an overall system-level *adaptation* occurs.[12] This adaptation process is reciprocal to the extent that entities influence their environment while being simultaneously receptive to influences by others.

The complexity theory focuses on the aggregate behavior of the system. As the network entities interact and adapt, simpler higher-order structures emerge. This process is referred to as *emergence*. For example, the collective wisdom of a network, a market crash, or a supply chain network breakdown would be an *emergent behavior*.

Scholars argue that *self-organization* is behind the examples of emergence.[13] The region of emergence, also called the *melting zone of maximum adaptive capacity* or *flux*, lies between the *edges of order* and *chaos*.[14] As Per Bak in his pioneering book *How Nature Works* proposes:

> ...Complex behavior in nature reflects the tendency of large systems with many components to evolve into a poised, "critical" state, way out of balance, where minor disturbances may lead to events, called avalanches, of all sizes. Most of the changes take place through catastrophic events rather than by following a smooth gradual path. The evolution to this very delicate state occurs without design from any outside agent. The state is established solely because of the dynamical interactions among individual elements of the system: the critical state is self-organized. Self-organized criticality (SOC) is so far the only known general mechanism to generate complexity.[15]

Bak and colleagues demonstrate the *self-organized criticality* with an intuitive *sand pile model*. Once the sand pile reaches a certain height, the adding of grains of sand on top causes sand slides, avalanches, or cascading effects of arbitrary size (with power law characteristics). The sand pile topples at a *critical state*.[16] Thus, emergence occurs with large catastrophic events. Minor events interact with other minor events, leading to such extreme intermittent incidents. This behavior is called *punctuated equilibrium*, a state that is at the heart of the dynamics of complex systems.[17]

Self-organized criticality is also ubiquitous in social systems.[18] Bankruptcy cascades are shown to behave in this way.[19] For example, traffic and logistics systems become dynamically unstable as they exhaust their capacity. Once the system reaches a *tipping point* due to sudden and unexpected events, the system's overall performance decreases. Dirk Helbing suggests the following to prevent such unexpected breakdowns:

> Optimizing for full usage of available system capacity implies the danger of an abrupt breakdown of the system performance with potentially very harmful consequences. To avoid this problem, one must know the capacity of the system and remain sufficiently clear of it. This can be done by requiring to respect sufficient safety margins.[20]

There has been some research to explain collapses and major events in societies with self-organized criticality. Jared Diamond, in his book *Collapse,* examines the collapse of ancient societal systems such as Maya, Anasazi, and Norse Greenland.[21] Gregory Brunk shows that as societies grow and mature, they harden and become rigid, and the complexity of a society—which is measured by the interactions between individuals—increases simultaneously. [22] This complexity, if it reaches a self-organized critical state, may cause unexpected major collapses: so-called complexity cascades. At such a tipping point, the system changes its regime and a *phase transition* sets in, also known as *bifurcation.* To avoid such major collapses, small external disturbances directed toward the system can be used. One should take mitigating actions to reduce societal complexity by allowing small setbacks to occur. These setbacks work to dampen system shocks. In sum, large cascades can be mitigated by small deliberate fluctuations. The very same idea is also used in wildfire management. Not leaving small fires to burn can lead the system to evolve into a tuned state, which in turn can cause much larger if not catastrophic fires to erupt.

The type of *coupling* in supply chain networks is also a differentiating feature. A network can be *loosely* or *tightly coupled.* Different supply chain networks achieve success with particular coupling structures. For example, movie and construction supply chain networks are more loosely coupled with many changing collaborators and alliances. In contrast, a tightly coupled network can be found in automotive supply chain networks. An instance of this coupling is Toyota's close-knit supplier network where long-term strategic alliances are maintained.[23] AnadoluJet, discussed in Chapter 10, displays the features of a tightly coupled supply chain network where a fixed, homogeneous, as well as highly synchronized flight network is managed. Table 3.1 displays the major differences between tightly and loosely coupled supply chain networks.[54]

3.2 Supply Chain Network Orchestration

> Your first and foremost job as a leader is to take charge of your own energy and then help to orchestrate the energy of those around you.
>
> **—Peter F. Drucker**[24]

The principal feature of a risk intelligent supply chain lies in its ability to integrate supply chain network stakeholders (i.e., suppliers, customers, employees, and partners). Supply chain networks need to be integrated along several dimensions so that they can operate as an integrated whole.

Think about the analogy of the glass. The value of an intact glass is that you can pour water in it and drink out of it. However, when that glass falls downs and shatters into pieces, there is no way you can drink the same water from any of those glass pieces. Even worse, you would probably cut yourself if you attempt to use any

Table 3.1 Tightly versus Loosely Coupled Supply Chain Networks

Tightly Coupled Supply Chain Network	*Loosely Coupled Supply Chain Network*
• Delays not possible in supply chain processes	• Delays are possible in supply chain processes
• Single method to achieve goals	• Alternative methods to achieve goals
• Little slack available in supply chain network resources (suppliers, partners, customers)	• Slack is available in resources
• Redundancies and buffers are deliberately designed in the supply chain network	• Redundancies and buffers are found spur-of-the-moment expediently
• Substitution of supply chain network resources and materials is limited and prethought	• Substitutions are improvised ad hoc
• Supply chain process sequences are invariant	• Order of supply chain process sequences can be changed if necessary

one of them. So, the functional value of an integrated glass disappears when broken. Now project this analogy to supply chain networks. A risk intelligent supply chain network is like an undamaged and intact glass. Once breakage occurs, individual glass pieces act as random items and have no value, except maybe to an artist who would put them together in a postmodern collage. It follows that the Integrator role of a risk intelligent supply chain is the first necessary step to move forward with the other roles of the I-Quartet model, that is, the *Inquirer, Improviser,* and *Ingenious.*

Moreover, the material valuation of a supply chain network depends on the stakeholders that participate in the network's operations. More than ten years ago, Kevin Kelly wrote about the value of networks in his book *New Rules for the New Economy: 10 Radical Strategies for a Connected World*: "In networks, we find self-reinforcing virtuous circles. Each additional member increases the network's value, which in turn attracts more members, initiating a spiral of benefits."[25] This feature of the networks is called the *increasing returns*, first introduced by Brian Arthur in the business world.[26]

Being an Integrator also requires acting like a maestro and orchestrating a supply chain network. There are successful examples of supply chain network orchestrators. A good example is the Li & Fung Company, which manages a network of 8,300 suppliers worldwide in 40 different countries. It manages garment and consumer goods for the best global brands of more than U.S. $8 billion value.[27] According to Li & Fung, network orchestration is concerned with both developing and managing the network and designing and managing specific supply chains through this network.[28] In Chapter 8, we will see that Kordsa Global is a unique case in global tire supply chain network orchestration. Kordsa Global manages its global production capacity in an integrated fashion linked to its global supply contract commitments.

Exploiting the orchestra analogy a little further generates even more viable insights. Just as the conductor leads and guides talented musicians to produce a good musical work, the network orchestrator also leads a group of talented suppliers to produce attractive products. Victor Fung, William Fung, and Yoram (Jerry) Wing describe the role of the network orchestrator in their book *Competing in a Flat World* as follows:[29]

- First, the musical director selects the music the orchestra will play. Likewise, the network orchestrator selects the projects and the best way to construct the supply chain to achieve that project. This requires understanding the opportunities for the customers.
- Second, the conductor decides on the types of instruments included in the group. Likewise, the orchestrator decides on the supply chain and the relevant suppliers to be included in that supply chain. Moreover, the tasks need to be determined to be assigned to specific suppliers. For this task to be effective, the orchestrator needs to have an intimate knowledge of the capabilities and skills of the suppliers.
- Third, as the conductor is in front of the group and stays on the same page, and in time the network orchestrator ensures the smooth flow of operations. Along the way, it solves the problems and suggests new solutions. At the right time, the right product at the right price should be on the shelves of the retailer. The respect of the network members is crucial for the orchestrator to continue to lead. Thus, winning this respect and creating an environment for the culture to evolve is necessary.

Similar to this discussion, C. K. Prahalad and M. S. Krishnan, in their book *The New Age of Innovation*, argue that in contemporary economics, surplus value is created for each individual by using global access to resources.[30] You focus on one customer at a time yet bring enormous amounts of know-how and know-why from a global pool of resources. While touching this global pool, interdependencies need to be carefully identified and managed, which will be discussed next.

3.3 Risk Interdependence in Supply Chain Networks

> The challenge of interdependent risks derives from network interactions. It is, therefore, not surprising that efficient solutions to these problems of interdependency require understanding and harnessing the power of the network.
>
> **—Howard Kunreuther**[31]

It is clear that any type of supply chain network risks, such as power shortages, labor strikes, major accidents, natural disasters such as floods and hurricanes, and severe congestions along transportation routes, will adversely affect the ongoing production and fulfillment of worldwide customer demand. According to Emery Roe and Paul R. Schulman, in the August 2003 blackout in the northeastern United States, financial damage was estimated at over US$6 billion.[32] As this example illustrates, the breakdown of a critical infrastructure can have cascading effects in a networked structure.

Sometimes, disruptive events can even occur interdependently; one event can easily increase the chance of another's occurrence. In those cases, losses can be immense. For instance, flooding may increase the risk of accidents for the distribution of commodities, which may simultaneously induce power shortages. For instance, in November 2006, a high-voltage line was shut down over a river in Germany for a ship to pass. This event caused a chain reaction of power outages that affected six European countries with 10 million people.[33]

Another example is Hurricane Katrina in 2005. When it hit the southern shores of the United States, not only did one of the worst human catastrophes occur, but also crucial oil, natural gas, and coal production was disrupted. These disruptions created cascading losses in more than ten critical industries. Total approximate economic losses due to this interdependence were calculated as US$5.1 billion in Louisiana for the first month after the hurricane.[34] Kordsa Global was one of the companies adversely affected in this hurricane as one of its major U.S. suppliers had to shut down production for two years.

These catastrophic examples clearly demonstrate that firms need to be cognizant of these subtle interdependencies across various disruptive events while managing their global supply chain networks.[35] What can companies and supply chain networks do to manage interdependent and cascading risks? It is clear that the more interdependent a supply chain network becomes, the quicker unexpected problems spread. Therefore, the degree of interconnectedness should be limited so that bigger problems do not occur.[36] In this respect, building firewalls in the supply chain network could easily help decouple the overall network from its affected parts in case of unexpected breakdowns. This strategy can essentially mitigate the spread of problems throughout the supply chain network.

Companies can also use a portfolio risk management approach. This can help assess the aggregate risk of a given supply chain network while still considering the various interdependencies across risk events.[37] This risk measure can then be used to compute potential supply losses within specific time periods. As markets and prices evolve dynamically, this risk measure can be updated considering these changes. Based on this aggregate risk measure, one can know how much of an operational loss is expected during a certain time period and eventually develop mitigation strategies—both operationally and financially—to hedge the supply disruption risk. Hedging strategies need to be evaluated cleverly when deciding on future investments.

3.3.1 Interdependent Risks, Weakest Links, and Tipping Points

How would a company invest in the management of a particular risk in the supply chain network? If there were no other stakeholders in the supply chain network, a company would decide the risk investment by itself. Yet the realization of a particular risk in the supply chain network depends not only on the particular investment decision of a company, but also on the investment decisions of its partners, suppliers, and collaborators in the network. Thus, one needs to speak of interdependent risk investment decisions.[38]

Let me explicate this concept within the field of airline network security. On December 21, 1988, a bomb exploded on Pan American Flight 103 from London to New York over Lockerbie, Scotland. Regrettably, 243 passengers and 16 crew members died. The bomb was arranged to explode at 28,000 feet. Interestingly, it was found that a bag with the bomb was first checked in to Malta Airlines in Malta with minimal security procedures. Then it was transferred to Frankfurt and eventually to London. Malta Airlines' weak risk management had negatively and indirectly affected the Pan American flight from London to New York. To put it in more general terms, the *weakest link* in the airline network determined the strength and resilience of the entire network. Even if your company invests in risk management, you pay dearly for another player's negligence upon which you depend. Hence, your risk investment buys less value than expected, sometimes even a negative one. So, what should be your strategy for risk investment decisions in such an interdependent network?

To understand an appropriate strategy, let us focus on external risks that supply chain networks are exposed to. External risks, discussed in Chapter 1, can be classified into two types: *man-made* or *intentional*, and *natural* or *unintentional*. Man-made risks include terrorist attacks, sabotage, theft, and espionage. Unintentional risks are the natural disasters such as earthquake, flood, hurricane, and storm.

Terrorists are more likely to launch attacks on less protected targets in the supply chain network.[40] Therefore, the probability of material and human losses depends on risk mitigation investments. As the investment level increases, the probability of losses will decrease. In supply chain networks, because the less protected targets are deemed more vulnerable to attacks, risk mitigation investments can be expected to rise.

On the other hand, there are negative externalities in natural disaster risk management. For instance, if a company does not invest in the earthquake risk management of its facilities and an earthquake hits, other companies in the same region will be adversely affected due to the rubble of the noninvesting company clogging the quake region. Hence, as the number of companies that do not invest in risk management increases, the general incentive to invest in risk management decreases for the entire supply chain network.

In a supply chain network, if one company or a group of companies changes its investment strategies for risk management, the equilibrium may radically shift. Even though, theoretically, no company would benefit from investing in risk management, once one company changes its strategy and invests due to external incentives or for some kindred reason, other companies in the supply chain network will follow suit. This kind of strategy switch is due to *tipping points*.[41]

Policies can be used to increase investment for risk management. Company by-laws can be used to increase the incentive to invest. Yet, in a supply chain network, rules may not prove effective. Thus, coordination mechanisms are necessary. Thinkable ways of ensuring coordination are as follows. First, supply chain contracts with flexible options can be used to coordinate behavior for investment in the network. Second, competencies can be delegated to higher coordination foci. Here the pioneering role of the industry and sector organizations may be helpful. For instance, the American Chemistry Council, an association of chemical manufacturers, plays the role of providing guidelines for firms for environmental, health, safety, and security performance of firms.[42] Third, in the supply chain network, selected companies acting as mindful leaders may set examples for the rest of the actors in terms of investment strategies. Last but not least, a risk and safety management culture is essential not only for individual companies, but also for the supply chain network as a whole. Here, with culture it is meant to have companies in the supply chain network firmly believe in joint learning, adaptation, as well as sharing of benefits and costs.

3.3.2 Earthquake Risk in Supply Chain Networks: The Turkish Example of Dealing with Earthquake Risk[43]

As a recent example of risk interdependence within the context of natural disasters, Rebecca Smith of the *Wall Street Journal*[44] stated that ninety-six nuclear reactors in the eastern and central United States are under an earthquake risk. After having experienced the cascading Japanese earthquake–tsunami Fukushima disaster in March 2011, it is imperative that not only potential earthquake but also tsunami risk should be considered for reactors lying near the East Coast of the United States.

As a consequence, interdependence as well as the cascade of risks has to be adequately considered. Failure cascades in networks require quick responses. If delays occur in responses, cascade spreading becomes inevitable even if large resources are used.[45] Moreover, risk perception biases have to be effectively managed for such cascading risks. When cascading events are in focus, people may underestimate the compounding power of multiple events. This risk underestimation is due to the so-called anchoring bias.[46] In a complex system, a low probability yet potentially extreme risk may be overlooked in short-term planning. However, such low probability risks may very well disrupt the complex system in its entirety (e.g., earthquakes, tsunamis).

The tsunami in Japan, no doubt, can be considered an extreme event—that is, a *black swan*—which only sporadically occurs. If one observes carefully, the frequency of extreme events is increasing with financial crises, extreme weather and climate change events, and global supply chain network disruptions. For example, in July 2012, according to the National Climatic Data Center's latest report, the United States has the largest moderate to extreme drought since 1950s.[47] During the same period, a similar drought has affected especially Turkey's central and eastern Anatolian regions by potentially reducing cereal yields.[48] Thus, anticipating and readiness to mitigate the effects of adverse black swans is a must for all private and public organizations, and surely for supply chain networks. I postpone the discussion on the peculiarity of black swans and their habitat, *Swan Lake*, to the upcoming chapters when we focus on the Inquirer and Ingenious roles.

In Turkey, the majority of the economic activity, including energy production and distribution, lies along the north of Marmara Sea, that is, the Sakarya–İstanbul–Çorlu geographical corridor.[49] This region has experienced major deadly earthquakes in its history, the most recent in August 1999. Figure 3.1 shows the time distribution of fatalities for the earthquakes, together with the four most deadly

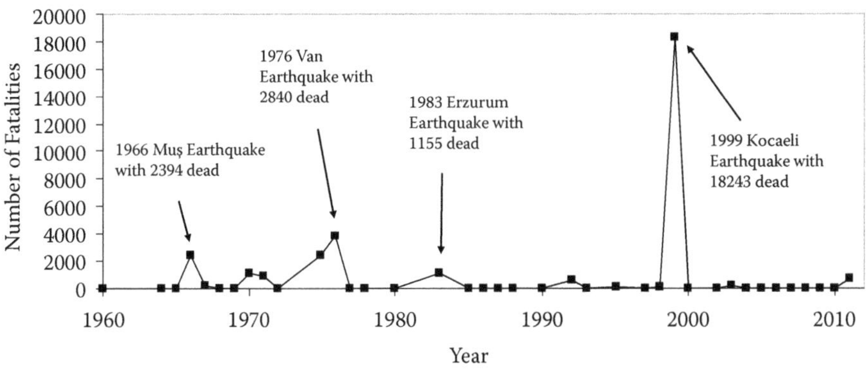

Figure 3.1 Number of fatalities in Turkish earthquakes during 1960–2011. (From http://www.afet.gov.tr.)

quakes during the same period. The 1999 Kocaeli earthquake was the most deadly among others during 1960–2011. In fact, Turkish people are unfortunately accustomed to major earthquakes and the resulting aftermaths (in fact, earthquakes have become a reality of life in Turkey). When you examine the natural disaster history of this country, earthquakes have created the highest number of fatalities, material damages, and economic costs. Table 3.2 sums up the total number of fatalities for the time period 1960 to 2011 brought about by major natural disasters, displaying the immense effect of earthquakes among other disasters.

With such a history of natural disasters, new facility, plant, depot, and customer service center locations must be planned in consideration of the new earthquake fault lines.[50] Yet proper risk assessment is still lacking in many organizations apart from a few risk intelligent ones, some of which represent the success stories presented in Part 3.

Can the mitigation of such known risks be neglected? If yes, at what cost? These are hitherto unexplored questions. Major highways and railway networks in this region are under the same threat. Not only individual locations but also the connections between these locations in the supply chain network are under threat. Once an earthquake occurs, humanitarian aid needs to be responsive. That is, humanitarian logistics operations require a sound, robust transportation network.[51] Problems in humanitarian aid response due to road bottlenecks were once again experienced in the recent 7.1 magnitude Van Earthquake in the eastern part of Turkey in October 2011.

Table 3.2 Total Number of Fatalities for Different Natural Disasters in Turkey during 1960–2011

Type of Natural Disaster	*Number of Fatalities (1960–2011)*
Avalanche	136
Earthquake	32,327
Flood	95
Forest fire	14
Heavy rain	4
Landslide	226
Mining accident	837
Rockfall	16
Other	246

Source: http://www.afet.gov.tr.

Each organization, private or public, has to answer the following questions: Have we correctly factored in potential external risks such as earthquake, tsunami, flooding, and hurricane into our strategic planning and supply chain decision making? Risk assessment of potentially interdependent risks is not a trivial task. Do we have the appropriate resources to conduct correct assessments? Answering these questions is followed by deliberation on finding the best available mitigation and hedging mechanisms against such risks. Insurance, ex post, can help in reducing financial losses. It does, however, not prevent human losses. Thus, besides insurance, one should consider operational hedging tools and techniques. One strategy could be to initiate sourcing from multiple suppliers (Refer to Chapter 1 for a discussion of multiple sourcing) based in different locations (uncorrelated or even negatively correlated ones with respect to particular supply chain risks). Thus, transportation possibilities as well as distribution patterns in the supply chain network must be carefully scrutinized when conducting such analyses.

3.4 Lessons from Turkish Carpet Weaving

> Carefully preserved in foreign collections as precious objects, several major carpet types, featuring the Turkic weavers' geometric gül medallions in distinctive configurations, became lastingly known by the names of European painters, such as Holbein and Lotto, who included them in their paintings, while the names by which the producers knew these carpets have been lost.
>
> **—Carter Vaughn Findley**[52]

In a supply chain network, to play the role of the Integrator, supply chain players need to look beyond the specific stage of the supply chain network where any type of glitch or disruption occurs to identify root causes. Often, root causes lie deep down in the surrounding links of the supply chain network. In fact, and to restate a previously made argument, the robustness of a supply chain network is contingent upon its weakest links.

I close this chapter with lessons from world-famous Turkish carpet weaving.[53] As a Turkish carpet is woven, the position of the weakest knot moves from one section of the carpet to another. Thus, in a dynamic supply chain network, firms should closely monitor the movement of the weakest spots and links—whether it is a customer, supplier, partner, complementor, or some other stakeholder—throughout the supply chain network and redesign their strategies accordingly to alleviate potential problems with these spots or links before they become the culprits of value drains in shareholder value.

Beginning with this chapter, for each I-Quartet role that follows, I provide a number of strategic questions any leader needs to ask and explore. Let us start with the Integrator role and following set of questions that could open up opportunities for a risk intelligent supply chain. Then, we move on to the next role, the Inquirer in the following chapter.

ARE YOU READY TO LEAD YOUR RISK INTELLIGENT SUPPLY CHAIN?

- How much intelligence do you have on the structure, topology, coupling, and features of your supply chain network (SCN)?
- Which player orchestrates your SCN? Can it be your company? Why? Why not?
- If you are the SCN orchestrator, do you manage the SCN considering the SCN robustness and complexity? Do you work toward decreasing the complexity and increasing robustness?
- Are you aware of the visible and latent risk interdependencies in your SCN?
- Which risks are the most critical to create cascading effects that need to be carefully managed?
- Do you manage risk mitigation investments jointly with other players in your SCN? If no, what is the value left on the table that can be recaptured?
- Do you dynamically monitor and discover the weak links in your SCN? Once you identify these weak links, how do you strengthen them to mitigate future disruptions?

Chapter 4

The Inquirer: Sensing and Learning in Swan Lake

An error gracefully acknowledged is a victory won.

—Caroline L. Gascoigne[1]

The most considerable difference I note among men is not their readiness to fall into error, but in their readiness to acknowledge these inevitable lapses.

—Thomas Henry Huxley[2]

The tragedy of life is not that man loses, but that he almost wins.

—Heywood Broun[3]

THE INQUIRER IN GLOBAL ECOSYSTEMS

Ants are extraordinary inquirers. They are constantly in motion learning from each other in the colony to be able to survive and thrive in any condition on earth. They are not discouraged from any setbacks and failures. They sense and anticipate potential risks and change their strategies for food search and nest construction.

THE INQUIRER ROLE IN BRIEF

The *Inquirer* role constantly senses and learns in the age of uncertainty and fragility. It intelligently learns not only from failures (small and large)[4] but also from near misses, precursors, and weak signals. The Inquirer role organizes its supply chain network to deepen its vision on the peripheries.[5] As it learns, it enhances anticipation capability for supply chain network risks.

4.1 Sensing and Learning in Supply Chain Networks: Brightening the Peripheral Vision

Luck may be something we create ourselves; it may not be due to the exterior universe, but I think there are people who have serendipity. Now it may be that they simply pay more attention.

—Nathan Kline[6]

To play the Inquirer role in a risk intelligent supply chain, first of all individuals, companies, and the supply chain networks must be effective learners.[7] Once a supply chain network gains the capacity to learn effectively, it is possible to develop the remaining capabilities of the Inquirer role.

Peripheral vision is the foremost capability in the role of the Inquirer. Peripheral vision is about sensing and seeing what lies around the corner; however, ironically enough, this attribute is not easy to cultivate as we often see only what we look for. Peripheral vision, unlike seeing as looking, involves exploring and discovering worlds beyond our cognitive and visual reach. Hence, sensing the fuzzy exterior zone, not the crisp core, is what creates the peripheral vision. In an executive survey conducted by George S. Day and Paul J. H. Schoemaker at Wharton and Insead, 80% of 150 global senior managers stated that their organizations' lack of peripheral vision has reduced the vigilance of their organizations in this age of speed and complexity.[8]

In the context of a supply chain network, peripheral vision is necessary in sensing weak signals from various stakeholders, markets, industries, countries, technologies, and competitors that the network interacts with at the peripheries. Surely, understanding the environment in which the supply chain network operates is critical to correctly interpreting the weak signals. The tension between overreaction and underreaction to these weak signals has to be managed with care.[9]

What is the main difference between focal and peripheral vision? Sensing signals from the periphery is not a trivial activity as the signals themselves are ambiguous and weak. Moreover, human beings and animals are genetically programmed to have focal vision. Hence, developing peripheral vision to sense, interpret, and learn from weak signals should be a deliberate strategy for organizations and supply chain networks. To this end, George S. Day and Paul J. H. Schoemaker, in their book *Peripheral Vision*, suggest seven steps to sense and learn from the peripheries as follows:[10]

- *Scoping*: Where to look
- *Scanning:* How to look
- *Interpreting:* What the data mean
- *Probing*: What to explore more closely
- *Acting:* What to do with these insights
- *Organizing*: How to develop vigilance
- *Leading*: An agenda for action

I next discuss risk sensing within the context of peripheral vision of a supply chain network. *Risk sensing* proposes a complex system to intuit the complexities associated with a particular environment. Chapter 5 discusses the response variety of a system and how it adapts to the environmental complexity within the role of the *Improviser*. To make sense of a variety of supply chain network risks, one should have a mind-set regarding similarities, differences, as well as subtle nuances about risks. In the peripheral vision of a supply chain network, risks can be classified based on their potential severity, likelihood, and their distribution in time. These supply chain network risks can be classified as follows:[11]

- ***Small versus large risks***: The potential impact of risks may differ. The risk of flooding can be large for a firm lying on the riverbed. Yet, breach of contract risk can be small for a firm that works with strategic partners. In addition, the size of the risk impact is also relative. A small to medium-size enterprise is not affected equally as a global company.
- ***Symmetric versus asymmetric risks***: The distribution of the potential gains and losses may be different. That is, in taking a certain risk, a potential upside might be higher than the potential downside. Thus, the willingness to take that risk could increase. For example, supplying a new market segment in a region could yield higher profits than remaining with the status quo. Yet this strategy could also evaporate the profits more quickly. How the distribution of the risk is shaped matters.
- ***Short-term versus long-term risks***: The impact of a certain risk may be felt in the short term. For example, when an earthquake hits a manufacturing plant, the immediate impact is on the production line. On the other hand, the deterioration of an overseas supplier's quality affects products much more slowly, with the likelihood that the financial impact is understood much later.
- ***Continuous versus discrete risks:*** Some supply chain network risks such as a strike hitting a particular supplier in the network greatly affect the flow of operations, whereas certain supply chain risks may be draining value continuously—sometimes latently—such as the exchange rate risk for a global company that operates on multiple continents. In that case, a supply chain network is exposed to continuous fluctuations in the exchange rate.

While sensing and interpreting supply chain network risks, *perception biases* as well as feelings and moods interfere and create noise in the background (and sometimes foreground). For instance, experts from various disciplines such as medicine, military, and academia are shown to be overconfident in their interpretation of events.[12] To have an effective peripheral vision, players in the supply chain network need to be aware of their perception biases.[13] Moreover, it should not be assumed that the risk perceptions of different players in the supply chain network would be similar. In fact, perception differences among players impact the quality of collaboration in risk investment and management.[14] Day and Schoemaker explain three major biases that a supply chain network needs to pay attention to:[15]

- **Selective perception**: This bias says that what we pay attention to depends on what we expect to see in the supply chain network and the environment. Paul Slovic says, "Strong initial views are resistant to change because they influence the way that subsequent information is interpreted. New evidence appears reliable and informative if it is consistent with one's initial beliefs."[16] Hence, whatever we do not expect to notice is filtered out of our peripheral vision. In this process, the context in which the weak signals are presented is also critical. In short, we might lose critical pieces of weak signals if we fall victim to this bias.

- **Rationalization**: Once we have a belief regarding a phenomenon, we tend to interpret the evidence we collect to support that belief. We might be in a state of wishful thinking neglecting evidences that counter our belief. Furthermore, this may create *overconfidence* in our belief. This bias is also related to the *fundamental attribution bias* where we put great importance on the actions of our supply chain network and discount the effects of environment.
- **Confirmation:** This bias says that we not only filter out information and interpret it with a distortion, but also seek evidence to bolster our perspectives. This bias may lead to a selective memory that conveniently forgets truths that do not fit into our preconceived world.

These major biases in a supply chain network may lead to underinvestment in risk mitigation for disasters such as earthquakes and hurricanes, as discussed in Chapter 3. Individuals' biases as well as feelings and moods (as well as the moods of societies as will be discussed in Chapter 6) may distort risk sensing and perception in the supply chain network. George Loewenstein, Elke Weber, Christopher Hsee, and Ned Welch succinctly explain the impact of feelings:

> Two dimensions are in action. One is dread, the extent of perceived lack of control and risk of the unknown, the extent to which the hazard is judged to be unobservable, unknown, new, or delayed in producing harmful effects. Studies found out that people in good moods make optimistic judgments and choices and that people in bad moods make pessimistic judgments and choices. Reading bad and good news also affects similarly. Fearful individuals make relatively pessimistic risk assessments and relatively risk-averse choices.[17]

The mood of individuals is also affected by their psychology and stress levels. Stress is shown to hurt the brain in a number of ways. John Madina sheds light on the detrimental effect of stress on brain functions and learning:

> Stress causes the body to produce a really nasty set of hormones ... glucocorticoids. They are good for short-term responses to trauma and strain, but they are not supposed to hang around for long. Certain types of stress can cause these hormones to overstay their welcome, and if they do, real damage occurs to the body, including the brain. The webbings between brain cells that hold your most precious memories can become disconnected. The brain stops giving birth to new neurons. Stress hormones seem to have a particular liking for cells in

the hippocampus, and that's a problem because the hippocampus is deeply involved in many aspects of human learning.[18]

4.2 Learning from Small Failures

Reward failure even more than success, and punish inaction.

—Robert I. Sutton[19]

The first section, which discussed peripheral vision, showed that collecting, interpreting, and making sense of information should be a major corporate function, especially in today's world where we have to live in data pollution. However, collecting more information does not necessarily lead to understanding and wisdom. Yoram (Jerry) Wind, Colin Crook, and Robert Gunther, in *The Power of Impossible Thinking*, shed light on the value of having more information:

> The effect of information in the past was to decrease uncertainty. Now sometimes the more information we have, the less we understand. Reports come from many sources, with widely varying characteristics, and we need to determine how reliable the information is. The interpretation is shaped by the agendas of the parties that are presenting it and receiving it. Rapidly changing information makes predictions about the future more difficult. A networked, nonlinear world of constant change—the global village—is also a turbulent world of transient fads and enduring truths.[20]

The first question we should answer is what type of information supply chain networks should collect and how this information needs to be interpreted to prevent major organizational and supply chain network failures. Failures (both avoidable and unavoidable) in organizations and supply chain networks are defined to be deviations from expected and desired results.[21] To prevent major failures, small failures need to identified and learned from. To this end, Mark Cannon and Amy Edmondson emphasize the value of learning from small failures:

> Our research in organizational contexts ranging from the hospital operating room to the corporate board room suggests that an intelligent process of organizational learning from failure requires proactively identifying and learning from small failures. Small failures are often overlooked because at the time of their occurrence they appear to be insignificant minor mistakes or isolated anomalies, and thus organizations fail to make timely use of these important learning opportunities.

> We find that when small failures are not widely identified, discussed and analyzed, it is very difficult for larger failures to be prevented.[22]

Learning from small failures can be sped up by *deliberate experimentation.* Design consultancy IDEO is a master of experimentation that promotes failing often to succeed sooner.[23] Cannon and Edmondson explain how successful organizations use smart experiments to learn from small failures:

> By devoting some portion of their energy to trying new things, to find out what might work and what will not, firms certainly run the risk of increasing the frequency of failure. But they also open up the possibility of generating novel solutions to problems and new ideas for products, services and innovations. In this way, new ideas are put to the test, but in a controlled context.[24]

Rather than inactivity, trying out new ideas though risky is essential to generate potential blockbusters. In fact, according to Robert Sutton, "Inaction is the worst failure, perhaps the only kind of failure that deserves to be punished if you want to encourage innovation.... Firms should demote, transfer, and even fire people who spend day after day talking about and planning what they are going to do, but never doing it."[25]

Creating the right environment is crucial for effective learning from failures. Edmondson, in a recent article in *Harvard Business Review*, emphasizes the importance of strong leadership in creating such an environment:

> Although learning from failures requires different strategies in different work settings, the goal should be to detect them early, analyze them deeply, and design experiments or pilot projects to produce them. But if the organization is ultimately to succeed, employees must feel safe admitting to and reporting failures. Creating that environment takes strong leadership.[26]

When a certain failure occurs in a supply chain network, a *postmortem analysis* needs to be conducted. For example, in the supply chain network, supplier failures may increase, and thus alternative sourcing mechanisms have to be used.[27] In such cases, several pieces of information need to be collected and interpreted. Why do these failures increase? Is it a unique event for a particular supplier? Or is it more widespread due to a structural change in the industry or market? How do the suppliers perform financially? Are bankruptcies also on the rise as suppliers fail in supply chain operations? Moreover, does our supply chain network experience events that we can call small or large failures as well as near misses (to be discussed next)? Have we prevented some supplier failures by introducing other measures such as providing trade credit to financially distressed suppliers?

Questions such as these and many more are necessary to identify the root causes of small and large failures as well as to find out the relevant precursors and near misses that can be used to anticipate and mitigate major supply chain network risks in the future. Now let us see how a supply chain network can learn from near misses and precursors.

4.3 Fail Fast, Learn Fast! Almost Fail, Learn Faster: Learning from Near Misses

> The real measure of success is the number of experiments that can be crowded into twenty-four hours.
>
> **—Thomas Alva Edison[28]**

Let us begin with a number of definitions. A *near miss* is defined as an event or series of events that could have resulted in one or more specified undesirable consequences under different but foreseeable circumstances but actually did not.[29] A near-miss event is also known as a *close call* or a *mishap* in different disciplines and industries. An *accident* is defined similarly as an event or series of events and circumstances that results in one or more specified undesirable consequences under foreseeable circumstances. A near miss is a special type of precursor. *Precursors* are defined as the conditions, events, and sequences that precede and lead up to accidents.[30] Thus, precursors can be considered the building blocks of accidents. When the necessary exacerbating factors are highly likely, the precursor is called a near miss.[31]

In practice, near-miss events are generally analyzed within the context of organizational safety and accidents in industries such as chemicals, petrochemicals, automotive, health care, and aviation.[32] However, near-miss management concepts are proposed to be useful in the context of service industries such as finance where operational risks are due to process, people, technology, and other external factors.[33]

Major accidents in different industries are generally rare and extreme and sometimes even catastrophic, such as those involving Three Mile Island, the *Titanic*, the *Challenger*, and most recently the Deepwater Horizon oil spill in the Gulf of Mexico. Due to the rarity of these accidents, learning has to be based on very few samples, sometimes on even just one data point.[34]

Extreme events are defined to have low frequency as well as high severity. Besides the major corporate accidents mentioned, weather events such as Category 5 hurricanes, typhoons, and earthquakes higher than 9.0 on the Richter scale are considered to be extreme events. There is a certain element of surprise in such events. Considering the element of surprise, these events can be classified as *counterexpected* and *unexpected events*.[35] Counterexpected events are defined as the events that have been imagined yet rejected as impossible. Unexpected events, on the other hand, are events that had never been imagined.[36]

Surely, managing unexpected event risks is not trivial. In our context, to be able to scientifically analyze extreme events in a supply chain network, we need a proper set of methods and tools. One useful candidate is the extreme value theory, which studies the statistical properties of extreme events in a variety of disciplines such as climate science, insurance, financial markets, and earthquake modeling. I need to note that people's worry and concerns in the supply chain network affect their effective use of these methods.[37]

Recently with the rise of societal impact of extreme events, more work is being done that combines a variety of disciplines. John Casti defines these extreme events as follows:

> These are events generally seen as deadly surprises whose likelihood is difficult to estimate, and that can turn society upside down in a few years, perhaps even minutes. X-events come in both the "natural" and "human-caused" variety. The asteroid strike illustrates the former, while a terrorist-inspired nuclear blast serves nicely for the latter. [38]

According to Casti, the extremeness of an event (X) can be computed by examining two statistics: the *unfolding time (UT)* (the time between the start of the event and the end); and the *impact time (IT)* (the period over which the event's effects can be felt).[39] He suggests that the extremeness can be measured with a formula: $X = 1 - UT/(UT + IT)$. Thus, if UT is short and IT is long, an extreme event such as a hurricane hitting Florida is more likely. When a diverse set of extreme events is being analyzed carefully, a number of fingerprints are identified. Casti summarizes these fingerprints as statistical, dynamics, evolutionary processes, more effect than cause, the only possibility of disaster and boom, and policy response.[40] A brief summary of these fingerprints closely following his work is as follows:

Statistical: Power law distributions (e.g., stable Levy, Paretian) are ubiquitous in the world of X-events. The frequency and the magnitude of events are measured with these distributions. Power law implies that an earthquake of twice as large magnitude is four times as rare.[41] Figure 4.1 displays a power law distribution (in log scale) together with the Gaussian (normal) distribution that has thinner tails.[42] According to Pierpaolo Andriani and Bill McKelvey, organizational dynamics and management research need to move to Paretian thinking as opposed to Gaussian thinking, where normality is assumed.[43] Power laws are ubiquitous in organizations. Power law is also observed in the positive and negative price movements of a variety of financial instruments.[44] Andriani and McKelvey state firmly, "Whereas normal distributions call for more standardized management, the long unique tails of rank/frequency Pareto distributions call for more unique managerial responses."[45]

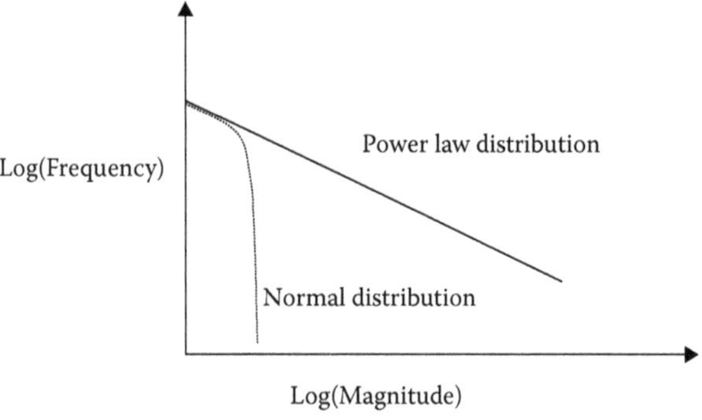

Figure 4.1 Power law versus normal distribution (log-log plot).

Dynamics: Systems with extreme events operate *far from equilibrium*. Thus, interaction of the system components and the collective effects become important in the behavior of the system.[46]

Evolutionary processes: System's evolution is determined by the extreme events. Thus, the system is being hit suddenly by the event and moved to a different evolutionary path. Hence, *path dependence* on extreme events is real.[47]

More effect than cause: Effects of these events are more significant than the causes. These events may cause large financial losses or fatalities.[48]

The only possibility of disaster and boom: Although the disaster risk could cause a poverty trap due to lack of credit in developing countries, a boom in technological innovation is also possible.[49] The possibility of such a positive extreme event (*Positive Black Swan)* is discussed in detail in the Ingenious role of the I-Quartet model in Chapter 6.

Policy response: In general, there is societal underinvestment in preventing extreme events. Costs are typically incurred after such an event occurs. Policy decision making should be trained to use probabilistic tools to address these events.[50]

4.3.1 An Example: Near Misses versus Major Supply Chain Accidents

Companies using near-miss information should normally gain benefits as the frequency of major accidents in their facilities goes down. In Figure 4.2, a chart displaying the major accidents and the near-miss event frequency of a global original equipment manufacturer (OEM) shows us that as the major accident frequency decreases in time, near-miss events tend to go up. This type of dynamics may have two interpretations. On one hand, there is an effective near-miss management. Major accidents are avoided at the near-miss level. Learning on the near-miss events reduced the likelihood of major accidents. On the other hand, the system produces

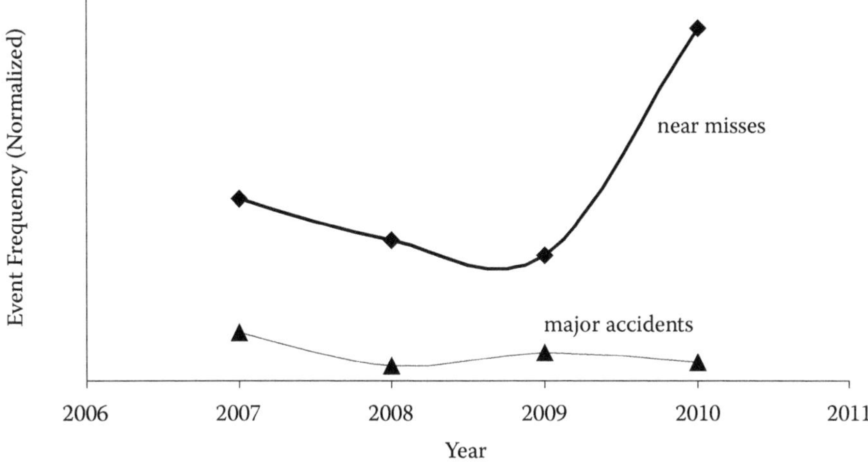

Figure 4.2 Major accidents versus near-miss events in time.

more near misses as time evolves. This also means that there is a higher risk of a major accident unless the event is caught at the near-miss level. Hence, system effectiveness has to be questioned and more carefully analyzed.

4.3.2 Near-Miss Intelligence in Supply Chain Networks

It is not difficult to see how near misses in a supply chain network can help us in anticipating and avoiding major supply chain disasters and disruptions. Near misses can be used effectively as *early warning systems* in supply chain networks.[51] Near misses are considered small problems and weak points in a supply chain network that need to be addressed soon before we experience a disruption.

An effective error and near-miss reporting and management mechanism is essential in achieving a risk intelligent supply chain, but this is easier said than done. *High reliability organizations* can serve as examples for emulation. According to Todd La Porte, "High reliability organizations exhibit extraordinary patterns of behavior and system performance across a wide range of varied and turbulent conditions."[52] And moreover, high reliability organizations show an unusual willingness to reward the discovery and reporting of error (even one's own error) without assigning blame.

We are aware that since major supply chain network risks are rare, sufficient data cannot be easily collected to conduct statistical analysis. Thus, collecting only major supply chain network risk data can be misleading for future predictions. However, using near-miss event data—which is potentially more frequent—in a supply chain network, one can refine the probability of risks causing major disruptions in the supply chain network. Because major disruptions are surprises for supply chain networks in general, they will greatly influence the risk perception when they occur. Dietrich Dörner says, "Catastrophes seem to hit suddenly, but in reality the way has been prepared for them." Thus, supply chain network catastrophes can be better anticipated if the near-miss events are carefully identified, gathered, interpreted, and acted on. I will detail the anticipation capability of a supply chain network later in the chapter. But now let us take a walk in the woods to reach Swan Lake in secluded nature.

4.4 Swan Lake: A Habitat of Colorful Swans

> The evening advanced. The shadows lengthened. The waters of the lake grew pitchy black. The gliding of the ghostly swans became rare and more rare.
>
> **—Wilkie Collins**[53]

After the 2008 world economic crisis, discussions in business and financial circles focused exclusively on black swans, or low-probability/high-impact risks—extreme risks that can dramatically break down markets, firms, and countries. Surely, understanding, anticipating, and getting prepared for disruptive black swan risks are necessary. Extreme financial losses increased many companies' awareness of the unexpected and sometimes unthinkable types of risks in practice. Several mitigation measures have to be rethought and redesigned.

However, limited focus on black swans—neglecting other types of swans (*white* and *gray swans*) in Swan Lake because their impacts are smaller—is dangerous at best and fatal at worst. A supply chain network needs an effective and intelligent management of all types of swans that live in this Swan Lake. Thus, management focus should be on not just one set of risks in the supply chain network but on all of them. The following section discusses the specific problems of such limited focus on particular types of swans, whether white, gray, or black. Supporting this argument, in 1998 in a presidential address to the American Risk and Insurance Association, Stephen P. D'Arcy used a canine metaphor to suggest that organizations need to study the *whole dog*—the entire spectrum of the risky outcomes—and not just the dog's wagging tail.

Considering the full spectrum of color in the swans swimming in the lake, I have expanded upon the metaphor by creating a habitat for these swans, borrowing the famous Swan Lake image. In our lake, supply chain network high-frequency/low-impact risks are identified as white swans, supply chain network moderate-frequency/moderate-impact risks are gray swans, and supply chain network low-frequency/high-impact risks are black swans.

Hence, the Swan Lake metaphor broadens the black swan metaphor already accepted in practice for extreme risks. Swan Lake also reminds one of the famous

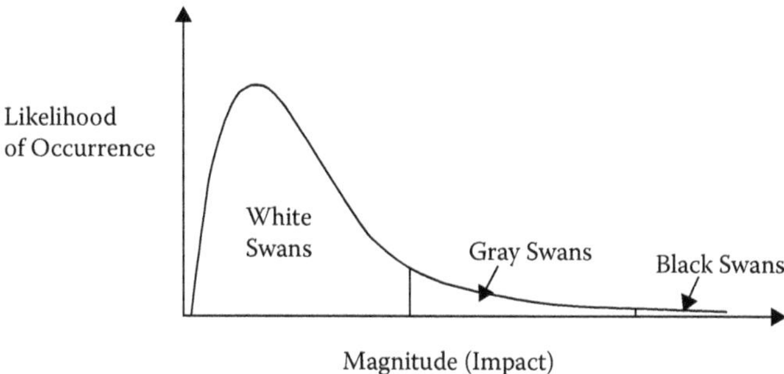

Figure 4.3 Swan Lake for global supply chain network risks.

ballet composed by Pyotr Ilyich Tchaikovsky. In this ballet, a swan image portrays faithful love, whereas in my metaphor it stands for various risks in supply chain networks.

In Figure 4.3, I illustrate Swan Lake with its three major birds and where they lie in a long-tail (power law type) distribution. The Swan Lake metaphor is also aligned with the complexity entailed in the supply chain network. As explained in the discussion of complex adaptive systems in Chapter 3, at a self-organized critical state, supply chain network risks occur at all frequencies and severities as demonstrated in Bak's sand pile model. To illuminate our visual perception on Swan Lake versus a black swan, Figures 4.4 and 4.5 display photos of a black swan and Swan Lake, respectively.

In Figure 4.3, white swans are located in the left tail of the distribution, where the impact of risk events is low and the probability of occurrence is high. Near-miss events are the gray swans in this context, appearing after the white swans. And lastly, black swans are located on the very right tail of the distribution. We are dealing with three types of swans, that is, three types of risk events in a supply chain network.

To be able to manage these swans in the lake, the first step is to understand how these swans behave. Of course, black swans are rare and thus almost nonexistent; therefore, we should brainstorm to discover the worst-case scenarios. Gray swans are the near-miss events that need to be analyzed as discussed. White swans are the easiest to learn from yet are easily discarded because of a false feeling of confidence.

How can Swan Lake evolve over time? First, the distribution of swans in the lake may change, which structurally changes the thresholds that separate the white, gray, and black swans in the supply chain network. A dramatic regime shift in the supply chain network structure may induce such a change. For example, we may acquire a firm that owns its distribution network in a certain region where we have no presence. We combine it with our existing supply chain network. Such a merger, unless managed carefully, can be detrimental for the overall health of the supply chain network as seen in the case of the Quaker–Snapple merger in the late 1990s,

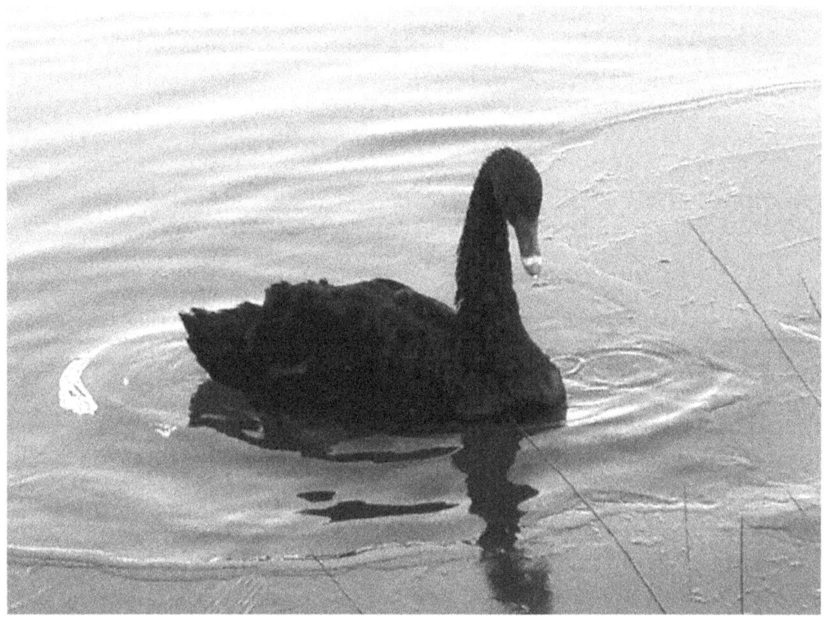

Figure 4.4 A black swan. (Courtesy of the author.)[79]

Figure 4.5 Swan Lake. (Courtesy of the author.)[80]

which caused Quaker to lose US$1.4 billion. Thus, the structure of the distribution of swans in the lake may change so that the potential impact of black swans increases. Hence, we cannot easily handle these birds.

On the other hand, in the absence of external shocks, as learning continues within the supply chain network among the stakeholders via positive constructive relationships, the distribution of swans in the lake changes such that the impact of potential black swans may go down and the behaviors of gray and white swans are better understood and internalized. In sum, our readiness to handle black swans improves as the intelligence of Swan Lake inhabitants increases.

4.5 The Ladder of Mastery: Swan Lake Ignorance to Swan Lake Intelligence

We are all born ignorant, but one must work hard to remain stupid.

—Benjamin Franklin[54]

Moving from Swan Lake ignorance to Swan Lake intelligence is a dynamic mastery process. During this process, there are intermediary states such as Black Swan ignorance and Black Swan anxiety that need to be managed well as they are not the final stops. Figure 4.6 displays the flowchart of this dynamic mastery process. WS, GS, and BS denote white swans, gray swans, and black swans, respectively. There are three flows in this process.

4.5.1 Swan Lake Ignorance to Swan Lake Intelligence via Black Swan Ignorance

You start to observe the Swan Lake as having no intelligence of any types of swans. That is, you are ignorant regarding the Swan Lake. Recognition of ignorance is a virtue and is the first step to gain more knowledge.[55] Hence, you first notice the white and gray swans in the lake. You observe their behaviors in the lake and gain intelligence on white and gray swans. In essence, you learn how to manage the Known Known, Known Unknown, and Unknown Known risks in the framework discussed in Chapter 1. Yet you are still black swan ignorant, which are the Unknown Unknown risks, even though these birds may be wandering in some other area of the lake. At this state, you become somewhat overconfident in managing risks. You believe that the residents of the Swan Lake consist of just white and

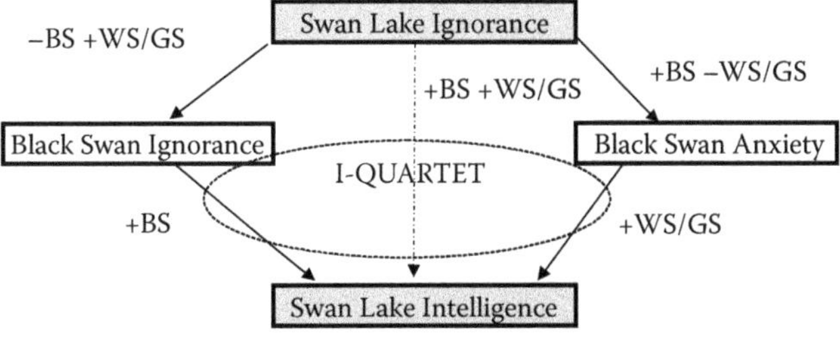

Figure 4.6 The ladder of mastery: Swan Lake ignorance to Swan Lake intelligence.

gray swans, nothing else. Overconfidence bias is dangerous and sometimes fatal. Your judgment of risks becomes impaired.[56]

After some time, one day finally you come across one black swan. It is a surprise for you. You become excited to have such an unexpected view. Then you begin studying the behavior of this black swan. Slowly, you increase your intelligence on both black swans as well as white and gray swans. This dynamic process requires the I-Quartet model roles, especially the Inquirer and Ingenious. Eventually, you become a master of Swan Lake Intelligence.

4.5.2 Swan Lake Ignorance to Swan Lake Intelligence via Black Swan Anxiety

In this flow, as opposed to the first one, you first encounter a black swan before having a chance to see the white and gray swans. You know that it is a rare bird. White and gray swans may be hiding in their dens or may have gone for a catch to some different corner of the lake. Observing the black swan first makes you a little anxious, fearful, and stressed because the impact of such a bird is much larger than the white and gray swans. I call this state *black swan anxiety*. Sometimes it is worsened by the stigma of black swan risks.[57] You begin to focus on the rare and extreme risks (i.e., Unknown Unknowns), somewhat overlooking the white and gray swans. In this state, the supply chain network becomes extremely mindful about the black swans. Organizations that pay excessive attention to tiny signals and are in a constant state of fear are called hypochondriacs.[58] This state is unhealthy and unsustainable and needs to transform into Swan Lake intelligence. If you can keep your calm during this state, there is a high likelihood that you will soon see the white and gray swans. During this dynamic process, you use the I-Quartet model roles, especially the Inquirer and Ingenious roles. Once the white and gray swans are observed in the lake, you gain mastery over Swan Lake in its entirety and thus reach Swan Lake intelligence.

4.5.3 Swan Lake Ignorance to Swan Lake Intelligence

This flow is the most direct one, yet it is the most challenging, most difficult, and rarest. There is a very small chance that a supply chain network uses this ladder of mastery. The rarity of this mastery ladder is because all types of swans (i.e., white, gray, black) are difficult to observe at the same time. Nevertheless, it is possible for some lucky and already prepared supply chain networks. What happens in this flow? As will be discussed in the I-Quartet role *Improviser* in Chapter 5, a supply chain network needs to have a sufficient level of internal complexity to handle the environmental complexity. If the supply chain network complexity is less than the complexity of the environment, fragility is imminent and extreme events are due. A supply chain network can be ignorant of Swan Lake, yet it can be open to different stimuli. It also may have the internal complexity and capability to sense, perceive, and learn. In addition, it has to also look for all types of swans in the lake to be able to recognize when it sees them.

4.6 Risk Anticipation in the Supply Chain Network

> If you have not thought about it before it develops, you probably won't recognize it when it does.
>
> **—Charles Doswell[59]**

Another capability of the Inquirer role of the I-Quartet model is effective anticipation—that is, an early warning system—for risks in Swan Lake for all types of swans. If one would like to anticipate supply chain network risks, one should have an effective mechanism to examine and interpret a variety of distress in the supply chain network. This anticipation of supply chain network risks can be achieved by various methods and tools proposed in various disciplines from engineering safety, catastrophe theory, earthquake modeling, and epilepsy prediction to critical transitions in complex dynamical systems. This section aims to provide an overview of the most relevant concepts, to provoke questions, and to leave the rest to readers.

4.6.1 *Early Warning through Near Misses and Precursors*

The first step in developing a robust early warning system is to enhance a supply chain network's near-miss and precursor intelligence. These events can play the role of *leading indicators* for *supply chain network distress*.[60] Suppose there are two precursor variables you are closely monitoring in your supply chain network for a particular supplier: (1) delays in the supplier shipments, and (2) quality defects in shipments. These two performance metrics may be positively correlated. The number of quality defects may increase as shipments are delayed. They may be interdependent.[61] For example, if a supplier starts to deliver late, to shorten this time gap it can speed up the production process, which eventually may deteriorate the quality. Surely, the sooner you learn about these precursors, the better. In software development, finding bugs earlier greatly reduces the costs of overrun and redesign later. Says Microsoft executive Grant George:

> The cheapest bug in any manufacturing process is always the one found earliest. Specification inspections, just like our formalization of structured and peer reviewed tests and build verification tests, are all catching bugs as early as possible. About 10 to 25 percent of all late stage problems can be found (or avoided) by following this approach.[62]

Moreover, you may experience a near miss during this process. In failing to meet shipment commitments, a supplier's transportation vehicles may take foolish risks in driving recklessly. This irrational risk taking may result in a near miss or fatal accident on the delivery trip. These precursors and near misses may become early warning signals in the supply chain network about this supplier and the ongoing relationship.[63] Your expectations may increase that a major supply chain network disruption is due. Your likelihood of anticipating such a major disruption increases in this scenario as you learn from precursors and near misses.*

4.6.2 The Shivers Model of Supply Chain Network Sickness Anticipation

Avert the danger not yet arisen.

—Old Vedic Proverb[64]

In this final section of the chapter, I present a conceptual model called the *Shivers model of supply chain network sickness anticipation*.[65] The name is derived using the metaphor of an individual who is coming down with the flu. Such a person will show certain symptoms, one of which is the shivers the body develops to maintain body heat. After experiencing this a few times, the person understands that he or she has the flu as the body suddenly shows signs of illness. In anticipation of this oncoming sickness, shivering serves as an early warning mechanism.

This physiological response is what I propose for a supply chain network problem where a major disruption is about to occur if a selected performance metric enters the Zone of Shivers. This conceptual model is qualitatively illustrated in Figure 4.7. Of course, as previously discussed, the type of performance metrics to monitor and collect information depends on the particular type of sickness, disruption, or problem we choose to anticipate. These metrics surely will be different, for example, for anticipating supply chain network disruptions. For natural disasters such as earthquakes, we could monitor the highway and railroad quality in the earthquake region, whereas for supplier bankruptcy risk we could monitor the liquidity and the supplier's return on invested capital.

In Part 3, which discusses leading Turkish companies, we will see that a diverse set of key performance and risk metrics as well as indicators is used for different purposes in the wider supply chain network. For example, Brisa uses various financial ratios of suppliers to anticipate supplier bankruptcy risk. Kordsa Global, on the other hand, monitors trends in future materials to anticipate technological risks in its supply chain network.

---

* A brief summary of mathematical modeling details is given in the Technical Appendix of this chapter.

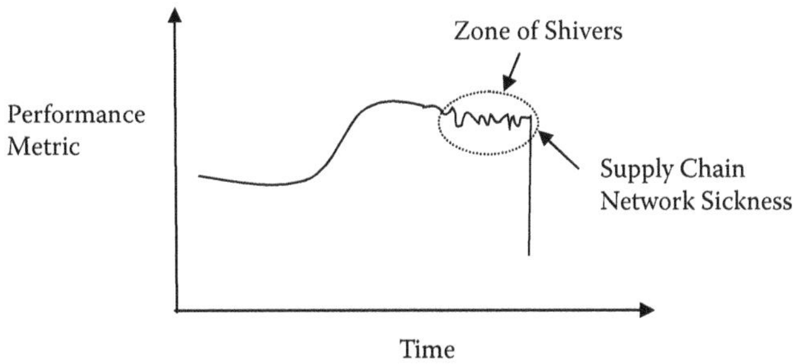

Figure 4.7 The Shivers model of supply chain network sickness anticipation.

The Shivers model is conceptually related to the theory of *critical transitions in complex dynamical systems,* such as biological ecosystems, financial markets, and climate. Critical transitions are defined as sharp transitions at tipping points to different regimes of operations in complex dynamical systems.[66] These transitions are abrupt and unexpected such as ancient climate shifts, collapse of ancient societies, and abrupt shifts of lakes from clear to turbid states.[67] Early warning signals are found in various systems to anticipate critical transitions. What follows are five possible early warning signals of a complex system that is approaching a critical transition.[68] The first three early warning signals are due to *critical slowing down* in a complex system.[69]

- **Slower recovery from perturbations:** As the system comes close to a critical transition, recovery from small perturbations slows down. One can conduct an experiment with the system to understand how close the critical transition is by applying small perturbations and measure the recovery rate.
- **Increased autocorrelation:** Besides the slow recovery from perturbations, if recovery rate is not easily measured in a system, then *lag-1 autocorrelation of fluctuations* in the system can be measured. Close to a critical transition, autocorrelation values tend to increase as slowing down makes the system state behave like before—in a correlated manner. For example, analyzing the climate change, a significant increase in autocorrelation was found.[70]
- **Increased variance:** Fluctuation variance is shown to increase when the system approaches a critical transition. As the system is perturbed and it slows down, each perturbation may result in a random walk pattern of fluctuations leading to higher variance.
- **Flickering:** *Flickering* is not due to critical slowing down as in the first three early warning signals. Flickering is observed close to a critical transition such that the system is constantly moving back and forth between two *basins of attraction* leading to a critical transition.[71] This type of signal has been observed in lake ecosystems where the lake state flickers between a clear and turbid state before going through a critical transition.

- **Increased spatial coherence:** This type of early warning signal is more relevant to systems where many agents interact among each other, such as financial markets and neural networks. This signal is particularly relevant for supply chain networks with interacting stakeholders. As the system is close to a critical transition, it tends to an increased *spatial coherence*, showing properties of increased cross-correlation as well as *resonance* among agents.[72] This state of resonance is related to the *synchronized state* and the occurrence of *dragon kings* to be discussed in Chapter 6.

Before I close this chapter, I need to mention an important caveat in using all of these early warning signals. Although these signals may serve as a warning in complex systems, predicting the actual moment of critical transition is difficult. Yet there is hope. Examining the financial market crashes, epileptic seizures, earthquakes, and material rupture models, Didier Sornette and colleagues offer suggestions for effective prediction of critical crashes of all sorts.[73]

Lastly, once a risk intelligent supply chain learns how to play the role of Inquirer, it is ready for the role of Improviser, where supply chain network resilience is achieved through creative–diverse–flexible tinkering. This is the topic of the next chapter.

ARE YOU READY TO LEAD YOUR RISK INTELLIGENT SUPPLY CHAIN?

- Do you work on honing your peripheral vision to detect risks lurking in your supply chain network (SCN)?
- Do you have risk sensing capability in your SCN?
- Are you aware of your and other players' risk perception and biases in your SCN? Due to mismatch of perceptions, how much loss do you think you incur that could be recovered by just changing the mind-sets?
- What does your Swan Lake look like in your SCN? Is it really full of white and gray swans with few black swans, or does it display a different composition?
- Do you have a formal mechanism to sense, report, and learn from small failures, near misses, and precursors? What is the award for reporting failures and near misses in your SCN?
- Do you assess your level of mastery in the Swan Lake intelligence ladder? Are you Black Swan ignorant, anxious, or intelligent? How far is your SCN to the top of the ladder (i.e., Swan Lake intelligence)?
- How do you anticipate potential critical transitions in your SCN? How proactive are you in anticipation of surprises? Do you have a formal early warning signal detection system in your SCN?

Technical Appendix

In a supply chain network, data can be collected for disruption and precursor events.

During this process, disruption frequency and disruption severity distributions need to be characterized, incorporating precursor events. Using these two distributions, financial loss distribution can be computed, which helps determine the right level of risk mitigation in the supply chain network.

Disruption frequency distribution: This distribution has to be estimated based on historical data. Data may be sparse if the disruptions are rare and extreme. *Bayesian updating* can be used to update the disruption frequency distribution by incorporating the precursor event data.[74] In this process, *false negatives* and *false positives*[75] have to be dealt with care. Bayesian updating processes must be also carefully designed as precursor signals used may bias the *posterior disruption frequency.*

Disruption severity distribution: Calculated in monetary units, this distribution helps determine the optimal level of risk mitigation for which a supply chain network should plan. Optimal levels of risk mitigation and insurance within a given computation can be addressed by a global risk management approach.[76] In other words, decision makers in the supply chain network have to optimally decide on the ex ante (risk mitigation) as well as ex post (insurance) strategies.

Financial loss distribution: This must be constructed using disruption frequency and severity distribution. *Worst-case loss, loss at risk,* and *conditional loss at risk* can be computed using VaR (Value at Risk) methods. This risk assessment methodology is used in market, credit, and operational risk frameworks.[77] A few years ago, we developed a risk measure called *supply at risk* to assess supply portfolio risk in supply chain networks.[78]

Chapter 5

The Improviser: Creative–Diverse–Flexible Tinkering to Achieve Resilience

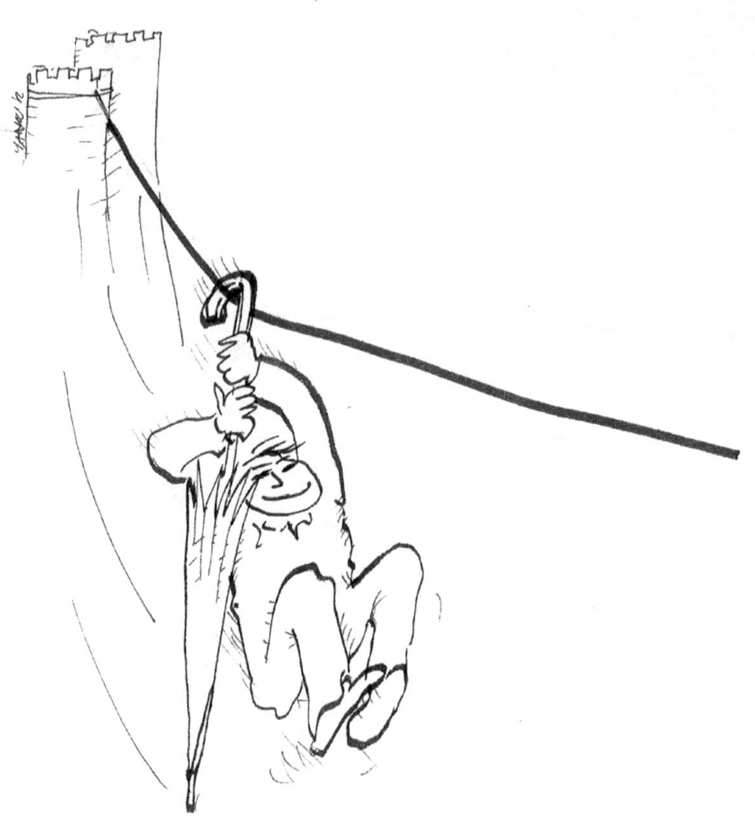

A caravan is arranged on its way.

—Turkish Proverb

Some of the most interesting innovations that happen, happen when the person doing it doesn't even know what's going on. True discovery, I think, happens in a very undirected way, when you figure it out as you go along.

—Joichi Ito, MIT Media Lab Director[1]

When fate hands you a lemon, make lemonade.

—Dale Carnegie[2]

To succeed, planning alone is insufficient. One must improvise as well.

—Isaac Asimov[3]

THE IMPROVISER IN GLOBAL ECOSYSTEMS

Honeybees have extraordinary improvisational capabilities. They search, tinker and collect the nectar from a diverse set of flowers in various environments and bring that nectar to produce delicious and nutritious honey.

THE IMPROVISER ROLE IN BRIEF

The *Improviser* role makes a risk intelligent supply chain more creative, resilient, and mindful under extreme periods of stress and uncertainty. In this process, the Improviser role creatively uses design thinking and oblique problem-solving approaches. Moreover, it tinkers and muddles through with the right level of diversity and flexibility to achieve supply chain network resilience.

5.1 The Improviser in the Supply Chain Network

> In the long history of humankind (and animal kind, too) those who learned to collaborate and improvise most effectively have prevailed.
>
> **—Charles Darwin**[4]

The third role in the I-Quartet model that a risk intelligent supply chain plays is Improviser.[5] Frank Barrett states, "Improvisation involves exploring, continual experimenting, tinkering with possibilities without knowing where one's queries will lead or how action will unfold."[6] Lee Tom Perry augments this:

> Improvisation is by no means a haphazard process; it should not be viewed as "anything goes" or "winging it." Instead, it should be accepted as a process governed by both freedom and form. The emphasis is on action and continuous experimentation, not obsessive planning.[7]

As improvisation includes continuous exploration, tinkering, as well as experimentation, Karl Weick says that "improvisation deals with the unforeseen, it works without a prior stipulation, it works with the unexpected."[8] Unexpected and unforeseen is what makes jazz a unique musical genre. Improvisation is the backbone of jazz. However, in jazz, spur-of-the-moment composition does not appear out of the blue. To address the criticality of previous experience, Paul Berliner states that improvisation in jazz requires practice, listening, and detailed study before performance. Thus, a jazz musician can be considered a highly disciplined "practicer."[9] Improvisation is also one of the crucial elements of Turkish classical and folk music.[10] In Turkish classical music, musicians improvise before, during, and after songs played in particular modes (musical modes are known as *makams* in Turkish classical music). This kind of improvisation is called *taksim* in Turkish classical music. On the other hand, similar improvisational embellishments exist in Turkish folk music where musicians open the path for a song, called *açış*, using the mode of the song to be played. Creative improvisation is one of the major manifestations of Turkish musical mastery that is obtained through years of hard and diligent practice.

Effective improvisation is required in turbulent and uncertain global supply chain networks. Today, improvisation plays a critical role in managing emerging risks especially under the threat of war and terrorism. Bob Ross of the Department of Homeland Security praises the Turkish navy in that respect. Being constantly alert, taking opportunities as they arise, and improving on the way makes the Turkish navy an effective improviser as opposed to the other navies of the world.[11] This is somewhat necessary per se. Operating in a fragile region with geostrategic significance (European Union to the west; Russia to the north; Central Asia, India, and China to the east; and the Middle East and Africa to the south) urges one to improvise.

When supply chain networks play the Improviser role intelligently, the resilience of the supply chain network increases. To achieve resilience, there are at least four crucial capabilities to develop, use, and continuously adapt as conditions change. In essence, supply chain network resilience is a function of these four capabilities:

- Creativity and design thinking
- Tinkering
- Diversity
- Flexibility

Resilience of a supply chain network is enhanced as creativity, tinkering, diversity, and flexibility improve, which are the fundamental capabilities of the Improviser role. In this part of the chapter, I will go over these capabilities in turn. Before I do that, I want to start with a discussion of the significance of resilience to set the stage.

5.2 Resilience

> More than education, more than experience, more than training, a person's level of resilience will determine who succeeds and who fails. That's true in the cancer ward, it's true in the Olympics and it's true in the boardroom.
>
> **—Diane L. Coutu[12]**

Resilience is addressed at individual, organization, supply chain, ecosystem, and societal levels in a variety of disciplines. Here, I provide a brief overview of ideas most relevant to this book. *Resilience* is defined as the ability of a system to recover to the original state upon a disturbance.[13] Moreover, resilience is about not only bouncing back from errors but also coping with surprises in the moment. Thus, to be resilient also means to use the change that is absorbed.[14]

At the individual level, resilience is also the ability to make do with whatever is at hand.[15] This skill is also called *bricolage* by French anthropologist Claude Levi-Strauss.[16] Bricolage is intimately connected to *tinkering*. Bricoleurs do tinker and improvise solutions with available resources. Diane Coutu describes the qualities of resilient people this way:

> Resilient people, they posit, possess three characteristics: a staunch acceptance of reality, a deep belief, often buttressed by strongly held values, that life is meaningful, and an uncanny ability to improvise.[17]

In her book *Uncommon Genius*, Denise Shekerjian follows the path of forty MacArthur Award winners and discusses the criticality of resilience in the lives of these highly successful people. She emphasizes the importance of focusing on the process rather than the goals in the process of building resilience, which is akin to the concept of tinkering and *oblique problem solving*. Shekerjian states for one of the award winners:

> By focusing on the process—bringing together the resources, meeting with people, learning from them, and seeing how their ideas improve her own—even if she never reaches her goal, she has succeeded. Her satisfaction comes from the doing, a fulfillment not experienced by people who focus only on winning and losing. Ignoring or undervaluing the process and lusting after victory makes it very hard to feel that your time has been well spent if you lose; focusing only on the goal makes it very hard to retrench and regroup and be resilient if you fail.[18]

At the level of *organizational resilience*, Karl Weick and Kathleen Sutcliffe list the three abilities of a resilient organization:

> The ability to absorb strain and preserve functioning despite the presence of adversity (both internal adversity, such as rapid change, lousy leadership, and performance and production pressures, and external adversity, such as increasing competition and demands from stakeholders).
>
> An ability to recover and bounce back from untoward events—as the system becomes better able to absorb a surprise and stretch rather than collapse [. . .].
>
> An ability to learn and grow from previous episodes of resilient action.[19]

If one moves from organizational resilience to *supply chain resilience*, one encounters a set of studies that present conceptual models of supply chain resilience. Yossi Sheffi, in *The Resilient Enterprise,* discusses the management responses of a resilient enterprise focusing on flexibility, redundancy, security, and collaboration.[20] In a more recent study, Timothy Pettit, Joseph Fiksel, and Keely Croxton define supply chain resilience as the fit between the supply chain vulnerabilities and capabilities.[21] Turbulence and changes in the environment create vulnerabilities. On the other hand, internal management controls create the supply chain capabilities. According to Pettit et al., as the managerial capabilities increase and external vulnerabilities simultaneously decrease, supply chain resilience increases.

At the ecosystems level, complex adaptive systems are studied with a so-called adaptive cycle framework. In this framework, dynamical systems such as ecosystems, societies, organizations, economies, and nations pass through four phases: (1) rapid growth and exploitation, (2) conservation, (3) collapse or release (creative destruction), and (4) renewal or reorganization.[22] In this cycle, major changes occur in the phases of collapse or release and renewal or reorganization. In the collapse phase, some components of the system may be lost. Yet in the renewal phase, novelties (new ideas, species, policies, industries) may appear. In the rapid growth phase, the ecosystem moves into a new trajectory. Later, the conservation phase creates a more complex, interconnected, and efficient ecosystem. However, resilience is reduced due to rigidity of the system and less variability. Then a creative destruction phase shifts the system into a different regime after crossing a threshold. And the cycle repeats itself.[23]

Considering the adaptive cycle framework, Brian Walker and David Salt, in their book *Resilience Thinking*, explain the criticality of resilience in ecosystem sustainability.[24] According to Walker and Salt, it is necessary in socioecological systems to be aware of *controlling (often slow-moving) variables*. In different ecosystems, slow variables include the following:

- Densities of sea urchins and sea otters in the Pacific Rim
- Nitrogen content of the soil on Easter Island
- Cover of seagrass in Florida Bay
- Grazing pressure as well as the level of rainfall in Australia's Northern Savannas[25]

System variables lie along these slow variables. *Thresholds* are defined as the levels in underlying slow variables of a system in which feedback to the rest of the system changes.[26] *Regime shifts* occur when the ecosystem moves beyond a threshold with undesirable and unforeseen surprises. If one knows the slow variables as well as where thresholds lie in an ecosystem, resilience can be measured by its distance from these thresholds.[27] Thus, one can increase the capacity to manage the resilience of the system by expanding the size of the desirable regime—which increases the capacity to absorb disturbances—to preclude a shift into an undesirable regime.

In a similar spirit, supply chain networks are complex adaptive systems that contain various slow variables, thresholds, and regimes. The resilience of the supply chain network has to be ensured for adaptability and sustainability. For example, the commodity price of iron ore may be a slow variable for the steel supply chain network, whereas high-tech industry clusters in China and India may operate as a slow variable for high-tech supply chain networks. To achieve a resilient supply chain network, a set of essential capabilities such as creativity and design thinking, tinkering, diversity, and flexibility should work together and are discussed next.

5.3 Creativity and Design Thinking

> Creative concepts often have a disruptive as well as a constructive aspect. They can shatter set patterns of thinking, threaten the status quo, or at the very least stir up people's anxieties.
>
> **—Kenichi Ohmae**[28]

A few years ago, Kjell Nordstrom and Jonas Ridderstrale argued in *Funky Business Forever: How to Enjoy Capitalism* that we live in world of *funk* where rules of the game are changing constantly.[29] Richard Florida, in his book *The Rise of the Creative Class*, took a similar stance in challenging perceptions when he said that the world is not flat as believed. According to Florida, the world is highly skewed in terms of production, creativity, and innovation.[30] People flock to different cities and urban centers to enhance and capitalize on their creativity.

In such a funky and nonflat world, you need creativity at the personal, organizational, and societal levels. And creativity is the first and foremost capability of the Improviser role in the I-Quartet model. As the late Steve Jobs stated:

> Creativity is just connecting things. When you ask creative people how they did something, they feel a little guilty because they didn't really do it, they just saw something. It seemed obvious to them after a while. That's because they were able to connect experiences they've had and synthesize new things.[31]

If you have always thought in the same way as you did before, you would probably end up with a similar set of solutions. Digging deep into the same surface to mine gold may not be a useful strategy where there are chances of finding gold in different locations if you dare to examine them. Most of the time, companies are complacent about moving away from the established norms and standards. Producing new ideas, models, and concepts that may actually help them move forward becomes futile. Uncertainties in global supply chain networks are so paramount that there is little probability that known methods even can yield feasible solutions. To this end, *design thinking* can be used effectively in supply chain networks. We should think about how we can design an idea, model, concept, tool, or technique from scratch. Design thinking is acclaimed as one of the key components of Apple's success.[32] At Apple, design thinking includes business experimentation, customer involvement, simplicity, and beautiful design.[33] Design thinking also requires *projective thinking* skills rather than *reactive thinking*. This apparent difference in thinking has been defined by Edward de Bono:

> In reactive thinking we analyze and sort the information that is presented to us. However, in projective thinking we, ourselves, have to generate the information and even create the context as we try to bring something about.... Reactive thinking includes problem solving and it includes judgment.[34]

Furthermore, *design thinking* is related to *opportunity seeking*. An opportunity is something you do not yet know that you want to do—and can.[35] Companies and individuals should intelligently seek opportunities. Surely, this search needs effective methods and *defocused attention*. Dean Keith Simonton, author of the book *Origins of Genius*, argues that "defocused attention permits the mind to attend loosely to more than one idea or stimulus at the same time, even when these cognitions and perceptions bear no obvious relationship to each other."[36] Opportunities are not only sought but are also spotted as well as built. De Bono makes the distinction between day-to-day problem solving and opportunity seeking:

> Problem solving also implies the removal of risk, whereas opportunity seeking implies increased risk and work. It is not at all difficult to see why problem solving is so much preferred to opportunity seeking. Management is forced to solve problems. No one is forced to look for opportunities until it is too late. By the time an organization is forced to look for opportunities it has probably already lost its best people, its market share, its credit rating, and its morale.[37]

The famous design consultancy IDEO is a master of opportunity seeking and spotting. Tom Kelley, general manager of IDEO, writes in *The Art of Innovation* about the work at his firm:

> We reject titles and big offices because they pose mental and physical barriers between teams and individuals. Paperwork is kept to a minimum. There's a friendly competition at work in the company. People grab projects and opportunities.[38]

I need to note that deliberate opportunity seeking also supports the *Ingenious* role of a risk intelligent supply chain where organizations look for positive black swans in Swan Lake. I will postpone discussion on the search for positive black swans to the Ingenious role chapter.

5.4 Tinkering

> I'm increasingly impressed with the kind of innovation and knowledge that doesn't come from preplanned effort, or from working toward a fixed goal, but from a kind of concentration on what one is doing. That seems very, very important to me. It's the actual process, the functioning, the going ahead with it.
>
> **—J. Kirk T. Varnedoe**[39]

To tinker is to fiddle, play, and meddle.[40] *Tinkering* is intimately connected to biological evolution where natural selection and recombination create novelties and thus adaptation and experimentation.[41] François Jakob describes the mode of tinkering operation as follows: "Often, without any well-defined long-term project, the tinkerer gives his materials unexpected functions to produce a new object.... Novelties come from previously unseen association of old material. To create is to recombine."[42] Recombination should be practiced for ideas, models, parts, components, and methods. For recombination to occur, one needs sufficient amounts of raw material to begin with. These raw materials, be they ideas, models, or parts, are the spare parts to construct novelties. In a risk intelligent supply chain, raw materials are composed not only of actual parts and components but also of various suppliers widespread around the globe. Using a diverse set of suppliers to create a specific product or service becomes essential in the competitive space.

In the context of a recombination process, Stuart Kaufmann calls the set of these first-order combinations *adjacent possible*.[43] He writes, "Today, the adjacent possible of goods and services is so vast that the economy, stumbling and lunging into the future adjacent possible, will only construct an ever smaller subset of the technologically possible."[44] Using this concept, Steven Johnson, in his book *Where Good Ideas Come From*, explains the process of bringing about creative ideas and novelty:

> Challenging problems don't usually define their adjacent possible in such a clear, tangible way. Part of coming up with a good idea is discovering what those spare parts are, and ensuring that you're not just recycling the same old ingredients.... The trick to having good ideas is not to sit around in glorious isolation and try to think big thoughts. The trick is to get more parts on the table.[45]

Thus, creative and good ideas depend on the quality as well as the diversity of spare parts. On the other side of the coin, the tinkering mind-set is contrasted with an engineer's mode of operation where a definite project and a set of tools are needed.[46] A similar dichotomy is also seen in the *methodism–experimentation* discussions in planning. Dietrich Dörner, in his book *The Logic of Failure*, explains:

> On the other hand, "methodism" . . . can impose a crippling conservatism on our activity. Many psychological experiments have demonstrated how people's range of action is limited by their tendency to act in accordance with pre-established patterns. To be successful, a planner must know when to follow established practice and when to strike out in a new direction. Recognizing the strategy appropriate to a particular situation—whether methodism or experimentation or some hybrid of the two—will help us plan more effectively.[47]

Tinkering is necessary to be an effective improviser that needs spontaneous novelty. Surely, one needs to be mindful during the tinkering process. Karl Weick and Kathleen Sutcliffe define *mindfulness* in their book *Managing the Unexpected* as "the rich awareness of discriminatory detail."[48] Mindfulness consists of various qualities. The quality of attention, seeing the big picture at the moment, concentrating on what is going on here and now, as well as capturing the anomalies and the unique features of the process can be listed as the most important.[49]

Harvard psychologist Ellen Langer first studied the mindfulness concept. According to her, being mindful consists of three key qualities: "creation of new categories, openness to new information, and awareness of more than one perspective."[50] Unless these qualities exist at the individual level, they cannot be practiced at the organization or supply chain network levels.

In the tinkering process, experimentation and adaptation are imperative where muddling through is practiced. In 1959, Charles Lindblom described muddling through as a process of decision making.[51] As opposed to a rational, direct, and comprehensive approach that evaluates all options at once (similar to methodism in planning), muddling through is an *oblique solution approach* that builds the decision step by step with constant adaptation and improvisation. Adding to this argument, Herbert Simon says, "Problem solving requires selective trial and error."[52]

John Kay discusses the oblique solution approach in his book *Obliquity*. According to Kay, "Obliquity describes the process of achieving complex objectives indirectly. . . . Oblique approaches often step backward to move forward" (Kay, 2011, p. 8). He further explains the oblique approach:

> The oblique solution of complex problems is a matter of managing the interrelationships between the interpretation of high-level objectives, the realization of intermediate states and goals and the performance of basic actions. Such skillful interpretation is required in even the simplest problem.[53]

The oblique problem-solving approach requires constant adaptation, discovery, and experimentation. To this end, we need *diversity* for the oblique approach and hence in tinkering. The next section explains why and how diversity is critical for a risk intelligent supply chain in the context of the Improviser role.

5.5 Diversity

> Diversity: the art of thinking independently together.
>
> **—Malcolm Forbes**[54]

> It is time for parents to teach young people early on that in diversity there is beauty and there is strength.
>
> **—Maya Angelou**[55]

Diversity is used in a variety of disciplines. Therefore, it is important that we have a working definition for it and know how it is used in global supply chain networks. I should begin with the fundamentals. It is defined as "a range of different things" in *The Oxford English Dictionary*.[56]

Operations strategy literature refers to the same concept using the *diversification* term. According to Hayes et al., diversification is accomplished with four patterns: (1) product diversification within a given market; (2) market diversification with a given product line; (3) process or vertical diversification; and (4) horizontal diversification (e.g., conglomerates).[57]

On the other hand, diversity is classified in biological ecosystems into three types according to Scott Page in his book *Diversity and Complexity*.[58] As supply chain networks are complex adaptive systems showing properties of ecosystems, this classification is especially crucial for us to understand and internalize the Improviser role of the I-Quartet model:

Diversity within a type, or variation: This refers to differences in the amount of some attribute or characteristic. The resilience of any complex system crucially depends on this type of diversity. This type of diversity is also called *response diversity* in ecosystems.[59] Response diversity plays the role of insurance for ecosystems. According to Marten Scheffer, ecosystems are composed of species called the *drivers* and *passengers*.[60] Drivers lead the way for an ecosystem, whereas the passengers have no effect in determining the direction. Driver species are similar in spirit to the supply chain network orchestrators. Simon Levin classifies the species in the ecological assembly as the *keystones* and *adamants*.[61] The removal of keystone species leads to the collapse of the integrity of structure. Also, the single keystone species' sensitivity to environmental changes can cause major shifts in the community composition.[62] Yet the adamants have minor effects. It is not always easy to determine which species play the role of keystone in an ecosystem. Levin states, "Predatory ants, for example, form a keystone guild in mountain birch forests, benefiting vegetation by reducing the role of herbivores."[63] So, as the resilience of an ecosystem relies on keystone species, keystone players in a supply chain network become the key in the livelihood of the network. The aforementioned supply chain network orchestrator such as Li & Fung may play this leading role, or a team of suppliers may take the lead as a guild. In Part 3, we will see why Kordsa Global is a keystone supplier in the global tire supply chain network.

Diversity of types and kinds, or species in biological systems: This refers to differences in kind, such as the different types of food kept in a refrigerator. This type of diversity is also called *functional diversity*.[64] This type of diversity creates *complementarity* between ecosystem species. This complementarity is obtained when a group of species is more productive than those in monoculture ecosystems. This concept is one of the pillars of *crowd wisdom,* which will be discussed shortly.

Diversity of composition: This refers to differences in how the types are arranged. Examples include recipes and molecules. This diversity is closely related to the modularity concept that is critical in risk intelligent supply chains. *Modularity* can be defined as the degree to which the nodes of a system can be decoupled into relatively discrete components.[65] In modular systems, one can use *buffers* against disaster cascades.[66] A laptop is a modular product. You may work on a particular component such as the hard drive independently without modifying the other components. If that component is damaged, others still work; hence, damage can be contained in a modular structure.[67] So, modularity enhances robustness. As seen in risk interdependencies, the system has to be well compartmentalized to prevent *risk cascades.* Table 5.1 displays the types of diversity and examples in risk intelligent supply chains.

Table 5.1 Diversity Types and Examples from Risk Intelligent Supply Chains

Type of Diversity	*Examples from Risk Intelligent Supply Chains*
Diversity within a type, or variation	Demand risk hedging by producing a substitute product in the same category
Diversity of types and kinds	Operational risk hedging with extra inventory or an extra reserved supplier capacity
Diversity of composition	Supplier portfolio diversity that includes a composition of a global offshore supplier with a local onshore supplier[100]

5.5.1 Diversity of Crowds

It is important to have diversity among stakeholders in supply chain networks. According to Scott Page, problem-solving and decision-making capacity is positively correlated with diversity in people's cognitive capability in any kind of group, such as a school, firm, or society. Diversity plays a role along four dimensions in the supply chain network stakeholders: (1) diverse perspectives, (2) interpretations, (3) heuristics (ways of solving problems), and (4) predictive models from diverse sets of stakeholders.[68] This enhances the value of the supply chain network and thus increases the supply chain risk intelligence. Problems are solved more effectively, risks are managed intelligently, and the profitability goes up.

People's different perspectives and their value in prediction are addressed in James Surowiecki's book *The Wisdom of Crowds*. He explores the collective forecasting capability of a group of people and compares it to one-man shows. He shows that a group of people's prediction capability is better under certain conditions.[69] Moreover, having a diverse set of people may preclude *groupthink* and *conformity*. Social psychology pioneer Solomon Asch conducted his famous conformity experiments in the 1950s. Maria Popova explains how conformity is broken with diversity of participants:

> While people slip into conformity with striking ease, it also doesn't take much to get them to snap out of it. Asch demonstrated this in a series of experiments, planting a confederate to defy the crowd by engaging in the sensible, rather than nonsensical, behavior. That, it turned out, was just enough. Having just one peer contravene the group made subjects eager to express their true thoughts.[70]

Multiple diverse sources of information, if effectively combined, are useful in getting a better estimate of the future failures, breakdowns, and disruptions in the supply chain networks. Learning from weak signals and precursors was discussed in the Inquirer role. Page presents the Crowd Beats the Average Law: "Given any collection of diverse predictive models, the collective prediction is more accurate than the average individual predictions."[71]

Information prediction markets are shown to be beneficial to elucidate the real wishes and desires of customers.[72] For example, the Hollywood Stock Exchange (HSX) is one such predictive information market.[73] It is shown that HSX performed well in predicting movie revenues released between March and September 2000. The prediction accuracy was close to that of a leading expert.[74] The recent trend in *crowdsourcing*—a method for distributed problem solving and production—is also shown to increase the effectiveness and speed of psychology experiments.[75]

In contrast, recent studies have shown that wisdom of the crowds is not always the case. Crowds can also become more ignorant than the individuals who compose them. More than merely ignorant, crowds can go crazy. Stock market bubbles and crashes are examples of such madness. If there is communication in the group, the diversity of the opinions may decrease as information flow affects people's perspectives. Conformity to a major opinion may take the lead, as seen in the Asch experiments.[76]

Nicholas Carr addresses possible *ignorance of crowds* in the context of *open-source innovation* by pointing to software programmer Eric Raymond's two models of open-source innovation: *bazaar* and *cathedral.* In the bazaar model, diversity of the individuals is crucial where peer production is the norm. Yet the cathedral model is based on the talent that creates the first seeds of the novel ideas. Carr explains the major differences:

> If peer production is a good way to mine the raw material for innovation, it doesn't seem well suited to shaping that material into a final product. That's a task that is still best done in the closed quarters of a cathedral, where a relatively small and formally organized group of talented professionals can collaborate closely in perfecting the fit and finish of a product. Involving a crowd in this work won't speed it up; it will just bring delays and confusion. The open source model is also unlikely to produce the original ideas that inspire and guide the greatest innovation efforts. That remains the realm of the individual.[77]

All of the ideas presented for the value of the diversity so far state that diversity increases the responsiveness of the supply chain network. *Responsiveness* has to be defined in a particular context, which is affected by the external *environment complexity.*

The Law of Requisite Variety by William Ross Ashby[78] helps us understand how a supply chain network can respond to external disturbances and stay resilient. This law states that the system has to have at least the same number of responses as the number of disturbances. In other words, you need an action—also regarded as a *conceptual slack*—to counter every perturbation or disturbance you are exposed to. If the disturbance diversity is high, the supply chain network also has to have a diverse set of responses.[79] To put it succinctly, you need to know more about what you do not know. For example, wildland firefighters are aware of the fire complexity and behave in a more complex manner to control the fires. They literally fight the fires with fires.[80] Casti states that the *complexity gap* between the internal system and the external environment is the major cause of extreme events.[81] Balanced *external vulnerability–internal capability,* discussed in the context of supply chain resilience, has interesting similarities to the complexity gap. Unless a supply chain network can enhance the internal capabilities, there is no way to give an effective response to the external vulnerabilities. Hence, unexpected—sometimes extreme—events occur.

When we deal with a supply chain network—which is a network of diverse sets of stakeholders—not only the response itself but also how it is being coordinated becomes critical. Because any type of response may not be effective to mitigate external disturbances. To this end, Yaneer Bar-Yam has extended the law of requisite variety to incorporate the coordination of components in a system.[82] Complex and uncertain environments demand decentralized organizations so that a variety of responses can be produced at different scales. Decentralized organizations have independent units that are loosely coordinated. In such systems, response variety is high. Yet when tight coordination among the system components is needed for a certain task, then during the process response variety is being reduced. Thus, there is a trade-off between the *variety* and *scale of response*. Bar-Yam puts it succinctly:

> Specifically, the Multi-scale Law of Requisite Variety implies that in order for a system to be successful its coordination mechanisms must allow independence and dependence between components so as to allow the right number of sets of components at each scale.[83]

5.5.2 Diversity–Redundancy Trade-Off

Redundancy of any components, parts, or players in a supply chain network increases the resilience. Redundancy is classified in two types: *pure redundancy* and *degeneracy*.[84] A pure redundant system contains multiple copies of the same part. If you have two smoke alarms in your manufacturing facility, when one fails the other is still functional. On the other hand, *degeneracy* means that multiple parts perform the same function or task but have different structures. In addition to a smoke alarm, an alarm based on heat becomes a degenerate option. Preserving redundancy is critical for supply chain network resilience. Yet there is a trade-off between diversity and redundancy unless the system capacity is unlimited. As you increase the diversity in a supply chain network, you essentially reduce the redundancy, and vice versa. You need the diversity for an effective response to a diverse set of disturbances of the environment. Yet you also need a certain degree of redundancy to prevent fragility of the supply chain network. Simon Levin emphasizes the criticality of redundancy in an ecosystem:

> Redundancy is the immediate source of replacement of lost functions.... The essential element to understanding the importance of redundancy is to elucidate the functional substitutability of one species for another, the ecological complement to economic substitutability.[85]

The redundancy–diversity trade-off is similar to the *exploit–explore* dichotomy in business strategy.[86] You need diversity to be able to respond to new disturbances, which requires exploration of new ways of operating. On the other hand, exploitation implies that you need to be better in what you are already doing so that you can capitalize on the already acquired responses. This is similar to having redundant parts to ensure continuity and robustness of the supply chain network. Resilient systems also have redundancy built into them.

Diversity is also shown to reduce the system volatility. Diversity can stabilize the fluctuations in the system via negative and positive feedbacks.[87] This helps maintain the *supply chain network stability.* Although ecosystem stability is intimately connected to the diversity, the linkage needs to be made explicit. Kevin Shear McCann sheds light on this issue:

> Diversity can be regarded as the passive recipient of important ecological mechanisms that are inherent in ecosystems. One promising mechanism that has been proposed recently is that weakly interacting species stabilize community dynamics by dampening strong, potentially destabilizing consumer–resource interactions.[88]

Interestingly, too much diversity may be detrimental to any complex system. Stuart Kauffman shows that in complex systems, as diversity increases, the number of conflicting constraints increases.[89] And this shifts the system from a *survivable* to *unsurvivable regime.* In other words, *system robustness* decreases abruptly after a certain threshold is reached for diversity. As we will discuss in Chapter 10, AnadoluJet is keen on limiting the diversity of its aircraft fleet—by believing in aircraft family, not aircraft type—to increase its network robustness. Figure 5.1 displays the phenomenon of this abrupt breakdown—a *phase transition (regime shift)*—as the diversity reaches a certain point. This phenomenon is also illustrated with an industry example regarding Unilever. As Kauffman puts it,

> Colleagues at Unilever noted to us that if they have a plant that manufactures different types of a product, say, toothpaste, then the plant does well when the diversity of products grows from three to four to ten

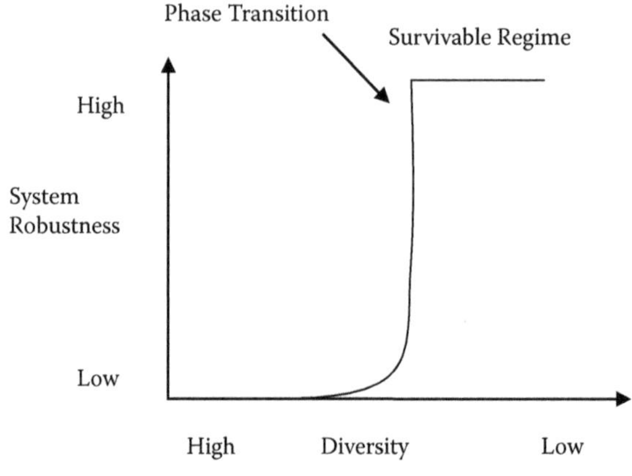

Figure 5.1 Transition from survivable regime as diversity increases.

> to twenty to twenty-five, but at twenty-seven different toothpastes, the plant suddenly fails. So rather abruptly, as product diversity in a given plant increases, the system fails.[90]

It is imperative that a risk intelligent supply chain be aware of these critical transitions and anticipate them before they occur. If a supply chain can stay on the survivable regime, then adaptation is possible where fragility can be prevented. To this end, the Improviser role has to work together with the Inquirer role.

5.6 Flexibility

> A man is born gentle and weak; at his death he is hard and stiff. All things, including the grass and trees, are soft and pliable in life; dry and brittle in death. Stiffness is thus a companion of death; flexibility a companion of life. An army that cannot yield will be defeated. A tree that cannot bend will crack in the wind. The hard and stiff will be broken; the soft and supple will prevail.
>
> **—Lao Tzu[91]**

I close the discussion in this chapter with flexibility. In supply chain networks, flexibility has been an important area of study for many years. My aim is not to provide a comprehensive perspective to this vast area but to focus just on the value of flexibility in achieving a risk intelligent supply chain.

Flexibility is the final necessary capability of the Improviser role for a risk intelligent supply chain. Specifically, flexibility is crucial in risk mitigation. Christopher Tang and Brian Tomlin show that five different flexibility strategies are beneficial in mitigating various (supply–process–demand) supply chain network risks. It is interesting to note that for every strategy demonstrated, a little flexibility—not too much—goes a long way toward risk mitigation.[92] These flexibility strategies and the types of supply chain risks that can be mitigated are as follows:[93]

- *Supply cost risk mitigation via multiple flexible suppliers*: A limited number of suppliers are sufficient to obtain the highest benefits in mitigating supply cost risk. For example, we will see that Kordsa Global and Brisa use dual or multiple sourcing strategies in their supply chain networks to mitigate supply cost risk.
- *Supply commitment risk mitigation via flexible supply contracts*: There are so-called *u-flexible* supply chain contracts that allow the supply chain players to modify their upstream orders based on the orders of the downstream. For a limited amount of flexibility in such contracts, the maximum value due to

flexibility can be obtained. Hewlett-Packard and Canon use such a supply contract, and the value of *u* is reported to be a few percent.[94]

- *Process capacity risk mitigation via flexible manufacturing:* In a manufacturing system, *h-flexibility* is defined as the number of *h* products a facility can produce. The higher this number for a given facility, the more variety that facility can produce. Again, with a limited amount of flexibility in the system, the maximum mitigation power for capacity risk can be achieved. *Full flexibility* is not necessary.[95]
- *Demand risk mitigation via postponement*: This strategy is shown to be valuable in mitigating demand risk and essentially postpones the time of *product differentiation* in a certain production line. As the time for postponement increases, flexible time to produce the generic products increases. With even a small amount of postponement, demand risk can be mitigated quite well. Surely, how demand evolves is critical in such a process and cannot be overlooked.
- *Demand risk mitigation via responsive pricing:* In this strategy, prices are decided on after more demand information is revealed. In other words, being responsive creates *timing flexibility* in a pricing process. As a firm delays deciding on the price even for a few periods, this timing flexibility mitigates the demand risk. AnadoluJet effectively uses responsive pricing strategy (see Chapter 10).

Flexibility is valuable not only for risk mitigation, but also for supply chain network orchestration. To this end, the Improviser role complements and enhances the value of the Integrator role. Li & Fung constantly asks itself what type of supply chain network it should design for different customer requests: *fixed* or *flexible*? In a fixed supply chain network, idiosyncratic investments are made to create a coherent product for a long time period. Complementary suppliers learn from each other and adding new partners inside the group is costly. In contrast, flexible supply chain networks come together for a specific project and are dissolved after the job is done. In a flexible arrangement, the orchestrator has to ensure the right quality and price of the final product.[96]

McKinsey & Company has also shown network flexibility to be of great value in reducing the manufacturing footprint where global resources are optimally used.[97] Moreover, operating a supply chain network in a wide geographic area creates its own flexibility as a real option. As difficult risks realize (such as currency exchange and demand risks), dynamically moving supply chain processes in such a network creates value for risk mitigation and hedging.[98] Kordsa Global is an interesting case for supply chain network flexibility, and will be further discussed in Chapter 8.

Building flexibility into organizational practices and supply chain network operations helps manage the diversity in the environment and markets. Flexibility increases the types and number of responses to a variety of stimuli. Yet companies easily might fail to understand the value of flexibility, according to Annelies

Vanderhaeghe and Suzanne de Treville.[99] They delineated nine rules at which companies fail at flexibility:

- Treating flexibility as a simple one-dimensional construct
- Not thinking about flexibility from the customer's viewpoint
- Thinking about flexibility only in terms of the product range
- Assuming that flexibility always results in a loss of efficiency
- Trying to increase flexibility in an environment where management and workers do not trust each other
- Trying to increase flexibility with unskilled workers
- Trying to increase flexibility while increasing capacity utilization
- Not considering extreme states of the world in defining the flexibility strategy
- Limiting flexibility initiatives to a single plant

Having discussed the key capabilities of the Improviser role, we are now ready to delve into the details of the final role of the I-Quartet model. The Ingenious role, building on the previous I-Quartet roles, performs the crucial intelligent risk taking in Swan Lake.

ARE YOU READY TO LEAD YOUR RISK INTELLIGENT SUPPLY CHAIN?

- Do you think that your supply chain network (SCN) is resilient? How do you measure SCN resilience? Do you constantly monitor the position of your SCN in the adaptive cycle?
- How much tinkering and oblique problem-solving approach are employed in your SCN? What can you do to promote their widespread use?
- How diverse is your SCN in terms of suppliers, partners, customers, markets, products, and services?
- Have you identified your *keystone* and *adamant* stakeholders in your SCN? What strategies do you employ to manage both of them?
- Are you aware of the diversity–redundancy trade-off? How much resource do you allocate to learn the interaction between the two?
- Do you tend to increase diversity without considering the increased complexity and possible phase transition to disruption? What levers can you use to prevent irresponsible diversity boom?
- What is your flexibility strategy in your SCN? Do you blindly increase flexibility without increasing its positive impact on various SCN components and metrics?

Chapter 6

The Ingenious: Intelligent Risk Taking in Swan Lake

If we expect the unexpected, we are better equipped to cope with it than if we lay extensive plans and believe that we have eliminated the unexpected.

—Dietrich Dörner[1]

Only those who will risk going too far can possibly find out how far one can go.

—T. S. Eliot[2]

And the trouble is, if you don't risk anything, you risk even more.

—Erica Jong[3]

THE INGENIOUS IN GLOBAL ECOSYSTEMS

Extremophiles are extraordinary ingenious organisms. They intelligently take risks and thus thrive in very harsh environmental conditions such as on cold ocean floors and polar ice, inside hot rocks deep under Earth's surface, and in desiccating conditions such as deserts.

THE INGENIOUS ROLE IN BRIEF

The *Ingenious* role takes risks intelligently. It not only mitigates supply chain risks with the best available tools and methods but also exposes the supply chain network to potential breakthroughs and blockbusters. Thus, it manages the two tails of uncertainty simultaneously with a delicate balance: It hedges against *negative black swans* and aggressively bets on *positive black swans.*

6.1 Surprises Are Not Surprises Anymore: Black Swans in Swan Lake

Life is singularly made to surprise us (where it does not utterly appall us.)

—Rainer Maria Rilke[4]

As discussed earlier in the book, the black swan concept was popularized by Nicholas Taleb in his book *The Black Swan*; he points out that rare and extreme events[5] occur as surprises, can be understood only after the events occur, and cannot be predicted. I have already discussed the basic properties of black swans in previous chapters.

In this chapter, I first differentiate between negative and positive black swans. Up to now, the focus was on the disruptive type of extreme events, that is, negative black swans. In this chapter, I also add to the discussion the other side of the coin: the constructive type of extreme events, that is, positive black swans, which create innovations, novelty, and blockbuster successes. Second, I focus on the positive black swans and provide two very different yet powerful perspectives—*positive deviance* and *extremophiles*—to better understand and nourish positive black swans in Swan Lake. Third, I discuss the ***dragon-kings*** first introduced by Didier Sornette to be able to study extreme events wilder than black swans. Fourth, I propose the *fork-tailed risk-taking strategy* that forms the backbone of the Ingenious role for a risk intelligent supply chain. Later, I complete the chapter by presenting the *closed-loop dynamics* of the risk intelligent supply chain considering all of the *I-Quartet* roles.

6.2 Negative versus Positive Black Swans

When one talks about extreme events as black swans, it is important to emphasize that a supply chain network is exposed to both types of extreme events, either disruptive or constructive.[6] Hence, black swans can be classified into two types: negative (disruptive) and positive (constructive) black swans. Negative black swans are the extreme event risks that disrupt the normal operations of a supply chain network. Any low-probability/high-impact risk that is detrimental to the supply chain network is a negative black swan. In general, a negative black swan could be a financial market crash, social unrest, an epidemic, or a material rupture.[7] In a supply chain network, any extreme breakdown of the flow of material, cash, or information in the network or any facility due to external or internal factors could be a negative black swan. In social systems—a supply chain network is one of them—where adaptive and evolutionary behaviors of agents are involved, negative black swans could be due to endogenous causes as well as exogenous shocks.[8] To this end, John Casti discusses the story of the impact of social mood on various extreme events in his book *Mood Matters*.[9]

On the other hand, positive black swans are the good, healthy, joyful, and happy events. They are the blockbusters, breakthroughs, discoveries, commercial successes, serendipitous learning, and innovations. As a recent example from sports, on August 5, 2012, at the London 2012 Olympic Games, Usain Bolt broke the Olympic record in the 100-meter race by finishing in 9.63 seconds and thus winning the gold medal. Interestingly, Bolt is a black swan in Swan Lake. Contrary to sports experts who thought that a tall man like him could not run as fast as shorter runners, Bolt appeared on the athletics scene as a positive black swan for himself and his fans and as a negative black swan for his competitors. In the end, he disproved the argument that tall men cannot sprint at phenomenal speeds and thus created an important black swan in sports history.

The movie and pharmaceutical industries, an unlikely pair, constantly search for blockbuster movies (hits) and blockbuster drugs, respectively. It has been shown that box office revenues and stock market returns of high-technology pharmaceutical startups show power law distribution properties with long tails.[10] This simply means that exceptional positive results are possible. According to Arthur de Vany and David Walls, in the movie industry a hit is generated by an *information cascade*.[11] Sequential dynamics of information discovery, transmission, and sharing create such a cascade, which then leads to a movie blockbuster. In Chapter 10, we will see that AnadoluJet deliberately seeks opportunities in the aircraft market such that positive black swans can be discovered. In the context of growths of countries, positive black swans may be considered the economic takeoffs. Mitchell Waldrop, in his book *Complexity,* quotes Stuart Kauffman:

> If all you do is produce bananas, nothing will happen except that you produce more bananas. But, if a country ever managed to diversify and increase its complexity above the critical point, then you would expect it to undergo an explosive increase in growth and innovation—what some economists have called an "economic takeoff."[12]

More importantly, one needs to be extremely mindful in the supply chain network during this process.[13] Recent neuroscience research sheds light on the social brain, arguing that "mindfulness requires both serenity and concentration, in a threatened state; people are much more likely to be 'mindless.' Their attention is diverted by the threat, and they cannot easily move to self-discovery."[14] Thus, rather than waiting for gifts of chance, active and deliberate exploration and searching for these positive black swans are necessary. In such exploration, accidental serendipitous discoveries play an important role.[15] Michael Michalko gives two very prominent examples from the history of science and innovation to illustrate this strategy:

> Alexander Fleming was not the first physician to notice the mold formed on an exposed culture while studying deadly bacteria. A less gifted physician would have trashed this seemingly irrelevant event but Fleming noted it as "interesting" and wondered if it had potential. This "interesting" observation led to penicillin which has saved millions of lives. Thomas Edison, while pondering how to make a carbon filament, was mindlessly toying with a piece of putty, turning and twisting it in his fingers, when he looked down at his hands, the answer hit him between the eyes: twist the carbon, like rope.[16]

The history of science is full of such examples. In 1879, Constantin Fahlberg accidentally discovered the artificial sweetener saccharin while working on coal tar. Charles Goodyear, while working on a rubber mixture, discovered the rubber vulcanizing process after dropping the sample on a hot stove.[17] On the other hand,

Herbert Simon states that unless knowledge and required experience already exist within the individual, accidents do not become surprises that can be easily capitalized upon:

> To exploit an accident—the image that appeared on Becquerel's photographic plate or the destruction of bacteria in proximity to the penicillium molds—one must observe the phenomenon and understand that something surprising has happened. No one who did not know what a dish of bacteria was supposed to look like could have noticed the pathology of the dish that was infected by the mold, or would have been surprised if it had been called to his or her attention. It is the surprise, the departure from the expected, that creates the fruitful accident; and there are no surprises without expectations, nor expectations without knowledge.[18]

In business strategy, formulating blockbuster strategies to deter imitation is critical in sustainable long-term success. *Blockbuster strategies* can sometimes be discovered as out-of-the-blue events. Jan Rivkin explains, "In complex settings, some very good strategies may go undiscovered for long periods. Thus a successful new way of doing business may emerge 'out of the blue' occasionally. Insightful and lucky managers may discover superior strategies—virgin peaks—even though tastes and technologies have not changed."[19] It is quite possible that while you look for something, you find something different but more valuable for you. To recognize such a different yet valuable strategy, you should be mindful.

Our main goal is to design and operate a supply chain network that is resilient, adaptive, and cascade resistant, as well as sufficiently diverse, that will effectively learn from near misses and weak signals. We should now look at what complexity theory teaches us to be able to understand the underlying dynamics. At the edge of chaos and order (also known as *flux)*, adaptation is possible to external environments via self-organization. Once the self-organized criticality is achieved, then all events show power law signatures. *Self-organized criticality* means that events exist at all magnitudes, mostly small ones with a few very large events, which creates, in our context, Swan Lake. Very few large-scale events become black swans. So to be able to nourish a supply chain network to create blockbusters and breakthroughs, we should be able to move the supply chain network to the edges of chaos and order, to the *emergence region*. In flux, the supply chain network will attain the self-organized criticality property, and then it will be able to produce potential blockbusters, which is also true for technological innovations that show avalanches (winds) of opportunity creation in the emergence region.[20]

The main question is*:* How can a supply chain network go into the flux region where discoveries can be made, and breakthrough results can be obtained? The flux region is between the order and chaos. Therefore, in the supply chain network, there should be sufficient levels of order as well as of chaos. Surely, each supply chain network has its own unique mixture of order and chaos, which can be discovered through sensing and learning as discussed in the role of the Inquirer.

For positive black swans to occur, diversity of the supply chain network is essential. Diversity creates the fertile soil from which positive black swans can germinate. As discussed in the role of the Improviser, diversity is one of the key components of supply chain network resilience. As the right level of diversity is attained in the supply chain network, supply chain resilience will increase, which also reduces the likelihood of negative black swans. So diversity has a dual role: increasing the likelihood of positive black swans and decreasing that of negative black swans.

As you have more diverse partners and opinions, chances will go up that you create interesting, remarkable, unique ideas or solutions in the supply chain network, thereby increasing the likelihood of positive black swans. For example, Scott Page argues in his book *The Difference*: "The application of new heuristics (in the form of skills and techniques) and perspectives (in terms of how to represent the problems confronted) drove growth in civilizations."[21] As we will discuss in Chapter 8, employee diversity in terms of culture, ethnicity, and background helps Kordsa Global manage its global supply chain network intelligently. In a more diverse supply chain network, the practice of *flying upside down*—that is, expanding thinking on outrageous events—is possible, which is useful both for imagining positive black swans and opening their paths to occur, as well as preparing for negative black swans.[22] Jerry Wind, Colin Cook, and Robert Gunther explain that flying upside down practice is used for airline pilots to educate them on "loss-of-control" crash scenarios.[23] Creativity and juxtaposition of diverse ideas create unusual and unexpected successes. Denise Sherkerjian puts forth the following statement made by one MacArthur Foundation Fellow Patrick Noonan:

> ... Patrick Noonan realizes that the future belongs to people who can integrate across disciplines, a sentiment that closely echoes the "surprising connections" definition of creativity, the idea of colliding frames of reference that produce an unexpected twist.... Bringing together a diverse set of talents to work out a problem from varying perspectives is how we will get creative solutions.[24]

Yet increasing the diversity haphazardly is a dangerous business. It has to be constantly monitored and dynamically managed. From complexity theory we know that as diversity increases, supply chain network resilience does not indefinitely increase. There is a maximum diversity, after which the supply chain resilience suddenly shifts to a much worse regime, as illustrated with the product diversity example at Unilever (see Chapter 5, esp. Figure 5.1). Diversity also brings complexity if not properly managed, and in complex systems unexpected events will inevitably occur.[25]

We discussed in the Inquirer role that intelligent learning from near misses and precursors is a must to mitigate future negative black swans. A supply chain network can observe near misses and precursors and also can sense the weak signals.

For negative black swans, these signals can be detected relatively easily and used for effective learning. Learning may decrease the complexity because surprising interactions can be reduced as experience grows.[26] On the other hand, for positive black swans, near-miss and precursor events are harder to characterize. We would not easily know that a near success would become a hit, yet we may use some leading indicators for potential positive black swans. One such indicator is the analysis of *bubbles,* which Didier Sornette describes this way:

> A bubble occurs when excessive political and public expectations of positive outcomes cause over-enthusiasm and unreasonable investment and efforts.... Only during these times do people dare explore new opportunities, many of them unreasonable and hopeless, with rare emergence of great lucky outcomes.... Bubbles are also times of self-organized self-excited auto-catalytic amplification of risks that allow the exploration of new niches.[27]

The boom of the railway in Britain in the 1840s, the Space Race and the Apollo program, the human genome project, and the information technology and Internet bubble in the late 1990s have been instrumental in creating unexpected breakthroughs and innovations.[28] Much more recently, a good example is Apple's iPhone, which would not have been a commercial blockbuster unless the boom of mobile handsets had not occurred in the 2000s. The explosion of technological innovations and global supplier production capacity investment led to Apple's iPhone breakthrough success.[29]

6.3 Positive Deviance: Identifying and Nourishing Positive Black Swans

> Indeed the capacity to be present to everything that is happening, without resistance, creates possibility.
>
> —**Rosamund Stone and Benjamin Zander**[30]

I now want to present the concept of positive deviance (PD), which is intimately related to the positive black swan concept. This concept simply states that in any system there may be people, ideas, models, and practices that create unexpected successes against all odds. If we can discover those rare successes, that is, positive deviants, we can then learn the underlying principles and further apply them elsewhere. There should be active search and discovery of the individuals or group of people who create these positive deviants because they often do not know what they know.[31] Once a positive deviant is discovered, essential learning is needed for the process of "how is it done?" not "what is being done?"

In their book *The Power of Positive Deviance*, Richard Pascale, Jerry Sternin, and Monique Sternin liken the positive deviance (PD) process to the work of nature:

> PD works like nature works. (This is not an analogy; it is the way it is.) Mutations in nature don't reinvent the whole genome of a species. Nature tinkers with a different shaped bird beak or slightly larger brain size that facilitates social intelligence. Natural selection does the rest, favoring variations that improve the access to food and reproduction. Of course, in nature, this all plays out in evolutionary time scales of centuries or millennia. Employing identical principles, the PD process achieves this change within months or a few years.[32]

The positive deviance concept was first successfully applied in health care to solve problems such as child malnutrition in Vietnam (65–80% reduction in child malnutrition in 22 Vietnam provinces) and female circumcision in Egypt (between 1997 and 2000, 4% drop in female circumcision).[33] Private companies such as Merck Mexico and Goldman Sachs also employed this approach, leveraged the positive deviants, and improved their productivity in sales and private wealth management, respectively.[34] Atul Gawande, one of the proponents of a positive deviance approach in medicine, provides five suggestions to become a positive deviant and make a worthy difference: "Ask an unscripted question, do not complain, count something you find interesting, choose an audience and write something offering your reflections, and look for the opportunity to change."[35] Through these suggestions, Gawande hopes that people become early adopters of change as well as be ready to receive interesting weak signals from the environment by asking exploratory questions and listening carefully. It is then possible to become a positive deviant to create novelty and value in the society.

How can we use the concept of positive deviance in supply chain networks? Positive deviants are the positive black swans that have already brought extraordinary successes to the supply chain network. However, they need to be discovered, understood, and disseminated throughout the network. To this end, we should first know that somewhere in the network a particular practice, strategy, policy, or a group of people has found a way to become adaptive to its environment and has been performing much better than the rest of the network. Starting with such a belief, we should then constantly look for these positive deviants, not overlooking any outliers.[36] The process is somewhat like searching for needles in a haystack. Nevertheless, the search must not cease. Once the needles are found, they will surely bring joy and happiness and sometimes wisdom to the whole supply chain network. Such positive deviants are widespread in extreme locations of biological ecosystems. This is the focus of our next topic.

6.4 Extremophiles: Positive Black Swans of Biological Ecosystems

That which does not kill us makes us stronger.

—Friedrich Nietzsche[38]

Extremophiles,[37] literally *lovers of extremes*, are biological life forms, single-celled organisms, microbes that can live and thrive in extreme environmental habitats of temperature, pressure, salinity, acidity, or alkalinity.[39] Deep sea vents, salt lakes, arctic waters, and acidic and alkaline lakes are some of the environments in which they choose to live in a resilient fashion. They are the positive extreme outliers in the world's organisms. Thus, I consider them the positive black swans or positive deviants. Bryan Appleyard summarizes the value of discovering and learning about these organisms: "Extremophiles also offer a cornucopia of new medical compounds, primarily antibiotics, as well as almost indestructible enzymes that could transform chemistry at both the domestic and industrial scales."[40] Many successful industrial applications are reported using extremophiles, such as production of enzymes for food processing, waste treatment, oil recovery, and laundry detergents.[41]

How about supply chain networks? What are the lessons from extremophiles that we could emulate to create a risk intelligent supply chain? Eileen Clegg suggests a set of general business lessons.[42] Here, I want to propose a set of particular lessons in the context of risk intelligent supply chains. First, in extremophiles, savior proteins are used in times of stress and disruptive changes that are mobilized for repair operations. Similarly, a supply chain network can identify its own savior protein, or *supply chain network healer*—be it a group of suppliers or a set of employees—which can withstand the external shocks and rectify the problems. Surely, these savior stakeholders need to be identified ahead of time, such that they are ready to fire when needed. Second, extremophiles can internalize external threats by intelligent mechanisms. For example, extremophiles living in highly saline water can dehydrate other organisms. In a similar fashion, a supply chain network after identifying external threats can internalize these threats. Being a collaborator with a competitor, thus creating a "co-opetition", is such an intelligent response.[43] Third, extremophiles also use the method of hibernation, bringing their metabolisms close to a halt. For instance, brine shrimp embryos can live in a habitat of no oxygen for decades.[44] A supply chain network can continue its operations at a very slow pace—without completely curtailing operations—if demand–supply risks go up suddenly. This slowing down may continue until conditions become more favorable.

Finally, extremophiles interestingly enough do not make use of the diversity of risk management strategies. In essence, they have a *niche* and *rigid response* to a particular environment. This ability is like having a monoculture as opposed to plural cultures that have diversity.[45] This type of risk management strategy is vulnerable if the conditions suddenly change preventing a timely adaptation. The role a supply chain network with a niche response strategy is a risky one considering the dynamic and increasing interactions with other supply chain networks. However, it is possible to have a portfolio of rigid responses that caters to specific niche habitats even in the same supply chain network. For example, a particular procurement risk management strategy in Asia may not be effective in Latin America, no matter what you do. Cultural peculiarities may strongly affect the type of risk management strategy you need to employ. Such an example is discussed in Chapter 8 where the Kordsa Global case is presented.

6.5 Dragon-Kings versus Black Swans

Supply chain network complexity depends on the *coupling (interaction) strength* of its individual stakeholders as well as their *heterogeneity,* among other things. Sornette demonstrates that as any complex system becomes tightly coupled and homogeneous, it will tend to move toward a situation in which *synchronization* and extreme risks occur, known as the *dragon-kings regime.*[46]

Dragon-kings in any complex system are defined as extreme events that are statistically and mechanistically different from the remaining smaller events.[47] The name dragon-king is given because these events show features of a different kind of animal: "dragon," which is a fantastical, superhuman extreme of great power; and "king," which refers to the fortunes of the monarchy.

Synchronization refers to the *coherent dynamics of coupled heterogeneous oscillators* in complex systems such as Per Bak's *sand pile model* discussed in the role of the Integrator.[48] In a supply chain network, these coupled oscillators can be considered the dynamically behaving adaptive players in the supply chain network that directly and indirectly interact through various flows (material, cash, and information) with each other. In this process, *positive feedbacks* play a crucial role. For instance, it is shown that in financial markets, positive feedback mechanisms such as imitation-based herding, portfolio insurance trading, and option hedging lead toward a dragon-king regime.

One can draw qualitative phase diagram of a supply chain network depending on the coupling strength and heterogeneity among supply chain network stakeholders.[73] In such a diagram, there are two major phases in which the supply chain network can operate: (1) self-organized criticality (SOC) (black swans) regime, or (2) synchronization (dragon-king) regime. As coupling becomes tight among the supply chain network stakeholders and these interacting agents become more homogeneous, it is likely to observe a synchronized regime where wild extremes, that is, dragon-kings, can occur.[49] Tight coupling is also shown to reduce the adaptiveness of complex systems to external changes.[50] Nevertheless, as the elements of your supply chain strategy become tightly coupled, this strategy grows more complex and thus more difficult to be imitated by competitors.[51] The strategy cannot be easily imitated; yet because tight coupling creates major time dependence in the system, one breakdown could trigger a series of other problems as demonstrated in an unfortunate 1996 Mt. Everest expedition.[52] On the other hand, if the supply chain network has some heterogeneity and the interaction strength is weaker, a self-organized criticality regime takes the lead, which produces black swan events alongside normal events.

6.6 Fork-Tailed Risk-Taking Strategy

All great reforms require one to dare a lot to win a little.

—William L. O'Neill[50]

Before I explain the fork-tailed risk taking strategy, which forms the major backbone of the Ingenious role, let me begin with risk-taking behavior in general. Risk taking is a strategic decision. Its short- and long-term implications do affect the sustainability of organizations.[54] Aswath Damodaran, in his book *Strategic Risk Taking,* states, "The key to good risk management is deciding which risks to avoid, which ones to pass through and which ones to exploit" (See Damodaran, 2007, p. 374). Furthermore, risk taking needs to be carefully calculated, as described by Herbert Simon:

> Creative performance results from taking calculated risks, where the accuracy of the calculation rests, again, on the foundation of superior knowledge. What appears to be a reckless gamble may be just that; but it's more likely that it was much less of a gamble than it appears, just because the risk-taker understood the situation better than competitors did.[55]

Taking risks intelligently is necessary in any type of endeavor, not only in business. To this end, Michael Schwalbe provides an interesting lesson learned in sports on risk taking:

> Many of the strategies employed in competitive and recreational sports are applicable in business and our personal lives. One lesson I learned from alpine ski racing was the "40-30-30 Rule." During training, early on, I tried to go fast, and I also focused on not falling. On a ride up the ski lift, my coach told me I was missing the point. He explained that success in ski racing, or most sports for that matter, was only 40% physical training. The other 60% was mental. And of that, the first 30% was technical skill and experience. The second 30% was the willingness to take risks. With ski racing, specifically, that meant taking the risk of leaning harder into turns, balancing at a steeper angle to the slope, and placing greater pressure on the outside ski edge—all of

> which increased the chance of falling. My coach explained, though, that if I wasn't falling at least once a day in training, I wasn't trying hard enough. Indeed, to improve at anything, we must at some point push ourselves outside our comfort zone. Body builders call it the "pain period." Only by trying something new, struggling, learning, and then trying again do we improve our performance. It's a simple matter of acclimating to unchartered territory.[56]

For some, risk taking is dancing with change, which is essential for new ideas. Joel Mokyr, discussing the technological progress of societies in his book *The Lever of Riches,* argues, "Willingness to bear risk is but one factor relating the intensity of innovation and uncertainty."[57] Denise Shekerjian quotes Debbie Meier as she talks about risk taking:

> I'm willing to take on risk because I'm invested in the project and because I believe in what we are doing here and want it to succeed. The drive is toward that, toward the goal, not the problems. Of course, I don't like to fail. I don't like to make mistakes. I'm afraid of looking foolish. I'm afraid of dying. But I'm not afraid of risk because risk is a part of change, and change is what new ideas are all about.[58]

Individuals should also be cognizant of their *risk appetite* before risk taking. Determining the right risk appetite—from risk averse to risk seeker—before a risk-taking decision is therefore critical.[59] Some studies have shown that gender affects risk aversion, with males more risk averse than females.[60] The Polish proverb, "Appetite increases as you start eating,"[61] may also describe risk taking. Once a supply chain network takes certain risks and does not get negative exposure for certain duration, risk appetite for those risks may inadvertently go up. A bias is formed, making the supply chain network believe that "it did not happen to us till now and thus will not happen from now on." Yet this *illusion of competence and control* in supply chain networks certainly increases the probability of getting hit by negative black swans. Also, overestimation is possible for the efficacy of the past methods if they were effective.[62] Moreover, successes are attributed to the skills and capabilities of the supply chain network partners rather than the role of chance.[63] In fact, this was one of the major reasons that led to the Chernobyl disaster. Dörner nicely puts it:

> If we never look at the consequences of our behavior, we can always maintain the illusion of competence. If we make a decision to correct a deficiency and then never check on the consequences of that decision, we can believe that the deficiency has been corrected. We can turn to new problems. Ballistic behavior [a behavior of having no influence on

> an action once it is launched] has the great advantage of relieving us of all accountability.[64]

On the other hand, for technological and creative progress in societies, overestimation of success by individuals may be beneficial such that risks are taken and opportunities used. Mokyr sheds light on this phenomenon, "If individuals consistently overrate their chances of success, their behavior may appear risk loving and society could enjoy more technological change relative to a situation in which individuals assess their chances correctly."[65]

To develop a viable, practical, and intelligent risk-taking strategy, one idea is to use the *maximin* type of approach.[66] That is, for a given negative black swan probability, we can maximize the probability of positive black swans in a supply chain network. Or we can minimize the probability of negative black swans given a certain probability of positive black swans. Using this approach, I propose two strategies that can be used in conjunction for intelligent risk taking.

The first strategy is the maximization of positive black swans for a given negative black swan risk coined *aggressive betting with smart hedging*. The second strategy minimizes the negative black swan probability for a limited positive black swan risk and is coined *conservative hedging with limited betting*. In the aggressive betting with smart hedging strategy, the decision maker bets on the upside potential in the supply chain network by hedging the negative black swans, that is, limiting the negative downside (disruptions in the supply chain network). In this strategy, a negative black swan risk is hedged up to a limit. Yet betting aggressively on the positive black swans is of higher priority. In contrast, a conservative hedging strategy implies that the decision maker minimizes the negative black swans for a given probability of positive black swans. Being conservative and preventing supply chain network disruptions are higher priorities than hitting the jackpot.

The fork-tailed risk-taking strategy combines these two strategies and uses them simultaneously. Rather than trading one for the other, the strategies are combined so the sum value of these two strategies is larger than the sum of individual values. Hence, there is a super-additive effect. Of course, one should be very careful which one to prioritize depending on the dynamics of the supply chain network. As conditions change, an aggressive betting strategy may be preferred to conservative hedging strategy or vice versa.

Table 6.1 displays a comparative list of various generic characteristics of supply chain networks (from functional strategies to network features) under two different risk-taking strategies. Of course, this table should be treated with caveat as it is only suggestive and just provides general approaches. Unique and specific nuances are always possible in practice. We will see that some of these specific approaches are used successfully by leading Turkish companies Kordsa Global, Brisa, and AnadoluJet, which will be presented in Part 3. For example, Brisa is keen on using

Table 6.1 Fork-Tailed Risk-Taking Strategy

Supply Chain Network Characteristics	*Aggressive Betting Strategy with Smart Hedging*	*Conservative Hedging Strategy with Limited Betting*
Outsourcing strategy	Near- or on-shore supply chain (focus and speed) and global outsourcing for diversity and creativity	Globally outsourced supply chain network (efficiency and cost)
Procurement strategy	Less long-term contracts, more spot purchases, and more strategic alliances for deep innovative capability and multiple and dual sourcing	More structured long-term contracts, less spot purchasing
Financial hedging	Options and Exotics	Forwards, Futures
Marketing strategy	Transnational marketing	Local, regional, and national marketing
Operational strategy	Agility, Flexibility	Efficiency
Market uncertainty	High	Low to moderate
Market complexity	High	Low
Risk features	Unknown/Unknowable	Known/Unknown
Geographic locations	Cities with high creativity index (e.g., Austin, Boston, Seattle)[68]	Cities with low creativity index (e.g., Louisville, Buffalo)
Mission	Market designer or maker since no market exists or vague	Efficient market player
Coupling in supply chain network	Loose	Tight
Decision autonomy	Decentralized	Centralized
Speed of cascading risks in supply chain network	Low	High

(continued)

Table 6.1 Fork-Tailed Risk-Taking Strategy (continued)

Supply Chain Network Characteristics	*Aggressive Betting Strategy with Smart Hedging*	*Conservative Hedging Strategy with Limited Betting*
Resilience	High	Low to moderate
Network topology	Scale free (emergent)	Random, structured, or vertically integrated
Learning potential	High	Low to moderate
Near-miss management capability	High	Low to moderate
Success versus failure focus	Success orientation (maximize successes and smart failures)	Failure orientation (minimize failures, content with small successes)
Failure tolerance	High	Low
Position of product or service in the life cycle	Fresh or new product and services	Mature product or service
Diversity in the human capital network	More diverse network (interesting backgrounds, ages, ethnicity)	Less diverse or heterogeneous network (similar backgrounds, experiences, age levels)
Product innovativeness	Innovative product or service	Functional product or service

a dual sourcing strategy in procurement such that hedging negative black swans as well as betting on positive black swans is possible at the same time.

6.7 Closed-Loop Dynamics of the Risk Intelligent Supply Chain

I'm a slow walker, but I never walk back.

—Abraham Lincoln[67]

A risk intelligent supply chain takes decisions using the different roles of the I-Quartet model. In simple terms, closed-loop dynamics illustrate different inter-

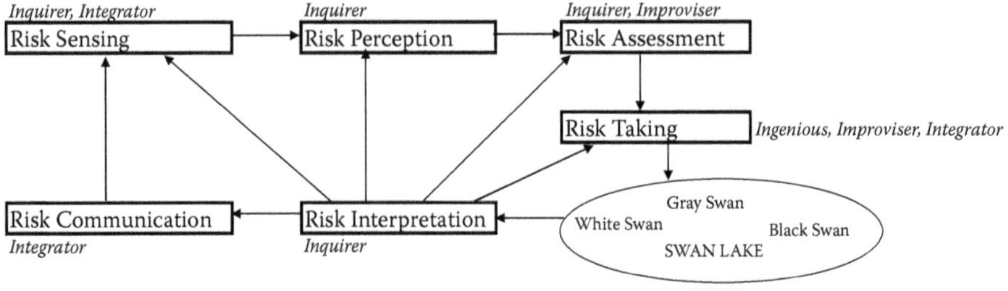

Figure 6.1 The closed-loop dynamics of the risk intelligent supply chain.

acting risk management decisions together with the I-Quartet roles leading these decisions. Figure 6.1 displays closed-loop dynamics.

A risk intelligent supply chain begins with risk sensing, which is managed by the Inquirer (and the Integrator). Once the risks are recognized, the right risk perception has to be incorporated by the Inquirer. In the third stage, risks are assessed with the help of the Inquirer and Improviser. Then, taking risks in Swan Lake occurs. Although the Ingenious role is the leader in this stage, it gets support from two other roles, namely the Improviser and the Integrator. Once the risks are realized in Swan Lake, they need to be interpreted with the Inquirer in the leading position. Lastly, the results must be interpreted and communicated effectively throughout the supply chain network.[68] This decision stage closes the loop and connects back to the first stage of risk sensing. In this closed loop, notice that the risk interpretation stage feeds information to all of the risk intelligent supply chain stages. Thus, this stage needs to be managed with somewhat more resources so as not to downgrade its speed, acuity, and accuracy.

6.7.1 An African Cheetah Looking for Food

Before moving on to Part 3, I want to wrap up this chapter and Part 2 by presenting a curious insight about the risk intelligence of the cheetah (*Acinonyx jubatus*), which is the fastest mammal, reaching speeds of 60 miles per hour (96 km/h) in pursuit of prey. Furthermore, the cheetah can go from 0 to 45 miles per hour in only 2.5 seconds.[69] However, this speed comes with its own fragility in that the cheetah becomes overheated after 20–60 seconds. This exhaustion sometimes causes the cat to lose its kill to scavengers.[70] Also, almost 50% of all attacks end in misery for the cheetah with total miss of the prey. A cheetah needs to gain energy and strength before the next attack a few days later.

Figure 6.2 illustrates the risk intelligent hunting of a cheetah wandering the African Serengeti savannah. Similar to a risk intelligent supply chain's closed-loop dynamics, a cheetah has to go through similar risk management decisions, some of them at once in short durations. Nevertheless, the story of a cheetah demonstrates the intelligent risk sensing, assessing, taking, interpreting, and communicating the results in a beautiful picture, even as beautiful as the cheetah itself.

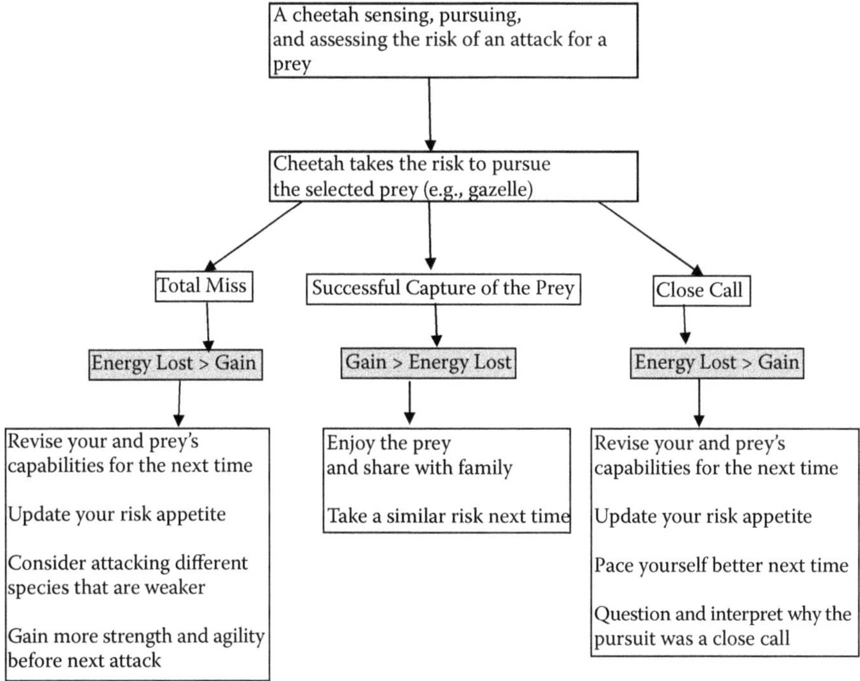

Figure 6.2 A cheetah's risk intelligent hunting.

ARE YOU READY TO LEAD YOUR RISK INTELLIGENT SUPPLY CHAIN?

- What are the potential and imminent negative and positive black swans in your supply chain network (SCN)? Is there a "boom" going on in your industry or some other industries that you may have overlooked that could potentially create black swans?
- Does your SCN promote and nourish positive black swans? Is your SCN in the region of emergence between edges of order and chaos? Can you identify your state in this complexity?
- Do you actively look for positive deviants in your SCN? Once you discover them, do you work on using their adaptive strategies throughout the SCN?

- What or who are the extremophiles in your SCN? Are they particular suppliers, partners, customers, or specific facilities? Can you emulate the risk management strategy of these extremophiles and make it widespread in the SCN?
- Besides black swans, are there possible dragon-kings in your SCN? Do you use a mechanism to imagine and discover those dragon-kings as well as determine your responses in case they occur?
- How tightly is your SCN coupled? Do you assess the tight and loose coupling as well as stakeholder heterogeneity frequently as your SCN evolves?
- Can you use a fork-tailed risk-taking strategy in your SCN? While employing such a strategy, will you tend toward aggressive betting or conservative hedging? Which of the two strategies will you use under what conditions? How should you decide on the balance of two strategies based on, for example, customers, markets, products and services, and regions?

3

HOW DO LEADING TURKISH COMPANIES THRIVE IN THE AGE OF FRAGILITY?

Chapter 7

Wisdom from Leading Turkish Companies

The tongue of a wise man lieth behind his heart.

—Ali Ibn-Abi-Talib[1]

A wise man hears one word and understands two.

—Yiddish Proverb[2]

Question and answer are both born out of knowledge.

—Mevlana Celaleddin Rumi[3]

Knowledge can be communicated, but not wisdom. One can find it, live it, be fortified by it, do wonders through it, but one cannot communicate and teach it.

—Hermann Hesse[4]

Part 3 contains three case studies of leading Turkish companies that manage global supply chain networks. My goal is to depict the specifics unique to the particular risk intelligent supply chain practices accomplished by Kordsa Global, Brisa, and AnadoluJet (a brand of Turkish Airlines).

Wisdom is defined in the *Oxford English Dictionary* as the "quality of having experience, knowledge, and good judgment" as well as "the body of knowledge and experience that develops within a specified society or period."[5] The meaning applied in relation to risk, however, is more akin to the latter because risk intelligent supply chain practices developed in Turkey are in our focus. However, as Hesse aptly observes, wisdom can neither be communicated nor taught, especially through writing. By naming, categorizing, and classifying—all rational, intellectual practices characteristic of writing—wisdom as a kind of truth is often eluded. Thus, instead of any sort of didactic analysis, in Part 3 I reproduce the lively discussions conducted with top Turkish business executives so interested readers and their firms can infer and interpret strategies of supply chain risk intelligence.

This chapter presents the broadbrush strokes of Turkish supply chain professionals' perspectives regarding risk perception and management in supply chains. The values inherent in these perspectives were evidenced by how these firms thrive and gain from the upside of the supply chain risks they intelligently take—thus necessitating this second look. The specifics of how resilient Turkish companies manage their supply chain risks constitute the discussions that follow.

At this point, I would like to emphasize the importance of the *idea transfer*. The ideas and approaches presented in this book emanate from leading Turkish companies. Yet their value lies not merely within themselves but in their ability to be transferred and applied effectively in other parts of the world. To this end, Edward de Bono mentions the importance of idea transfer as follows:

> Most executives live in the world of their own business and know relatively little of what happens in other areas. And yet it may happen that in one of those areas a procedure has had to be developed (because the need was greater there) which could be applied directly to another business area. Even if a procedure is not directly transferable it may serve to stimulate new thinking about an area.[6]

My hope thus is that the ideas expressed here travel not only outside of Turkey but also to the industries they represent.

After presenting the global tire supply chain in this chapter, I compare side by side the two most critical supply chain stages of the global tire supply chain, going from upstream to downstream: the tire cord and fabric producer Kordsa Global; and the tire manufacturer Brisa. All of these discussions with Turkish executives are self-contained. After presenting the two cases that operate at the upstream stages of the tire supply chain, I move to the dramatic case study of AnadoluJet, the newest and fastest growing brand of Turkish Airlines. I present this case while providing the historical evolution of the airline sector in Turkey.

Chapters 8, 9, and 10 are given in a specific sequence in the book. The sequence is designed by keeping in mind the flow from the upstream player toward a downstream player in the tire supply chain, as displayed in Figure 7.1. An individual using a car is a customer of an original equipment manufacturer (OEM)—such as Honda, Volkswagen, and Ford—which is a customer of Brisa for tires. On the other hand, it is interesting to note that a similar positioning for AnadoluJet—using Boeing aircrafts—also uses aircraft tires manufactured in this global supply

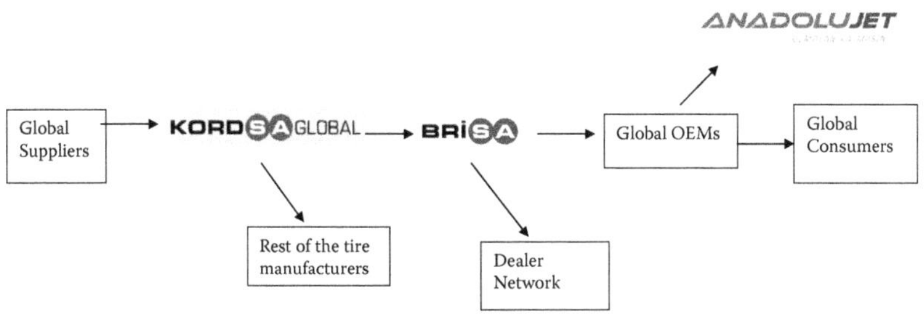

Figure 7.1 Global tire supply chain and position of the leading Turkish companies.

chain network. Thus, it is also located in one of the downstream stages of the global tire supply chain network.

I aim to present the most critical supply chain risks of these leading companies as well as the supply chain risk intelligence they honed to manage these risks successfully. Managerial lessons and insights are essential and can be emulated by other companies not only in the same but also other industries. Applicability of ideas and concepts especially in an emerging market context is of higher significance. Key lessons are derived and summarized within the context of the I-Quartet model after each chapter while considering an emerging and global market context.

In the chapters that follow, we will see that Kordsa Global, Brisa, and AnadoluJet all perform four essential I-Quartet roles. On the other hand, it is also important to state that some roles among the four are slightly more visible than others in each of the companies. First, Kordsa Global's greater strength in the Integrator, Inquirer, Improviser, and Ingenious roles enables it to orchestrate a truly transnational supply chain network. Second, Brisa performs much better as a virtuoso in the two I-Quartet roles, Inquirer and Ingenious. Sharing a joint venture with a global giant (Bridgestone) while serving a very competitive market urges Brisa to take risks much more intelligently as well as learn and discover constantly. Third, AnadoluJet's focus in setting up innovative methods to manage its airline network, taking intelligent risks, and improvising as it grows centers it more within the Inquirer, Improviser, and Ingenious roles.

Some remarks on the method of presentation of these cases are now in order. To decipher the gems of supply chain risk intelligence at Kordsa Global and Brisa, dynamic managerial conversations with top executives of supply chain, procurement, and contract management at Kordsa Global and Brisa are presented as dialogues. Key questions were used that opened up the lively discussions and thus created more avenues for the participants to speculate on further ideas and concepts. My goal in this structure is to make you, the reader, think and encourage asking more questions in your own companies and supply chain networks. As Claude Lévi-Strauss stated, "The scientist is not a person who gives the right answers, he's one who asks the right questions."[7] This is the advice I follow. This style of presentation was inspired by the *Platonic Dialogue*, where philosophical dialogues were given as conversations between characters.

In contrast to the Chapters 8 and 9, you will see that the dynamic conversation style is not preferred in Chapter 10's discussion on AnadoluJet. The historical evolution of an entrepreneurial idea (AnadoluJet) from its inception within the context of an established organization (Turkish Airlines) better suits an essay format presentation. To this end, I discuss not only how supply chain risks are being managed but also invaluable strategic lessons in building and operating a flight network, revenue management and pricing, marketing and sales, and customer service.

I now welcome you to enjoy the ride with thriving Turkish companies and their executives that have been managing their risk intelligent supply chains with vigor and vigilance.

Chapter 8

Kordsa Global: Orchestrator of Transnational Supply Chain Network

Profit is made while buying, not while selling.

—Sakıp Sabancı[1]

8.1 Prelude

The case of Kordsa Global in this chapter deals with the global tire supply chain and how it is being managed with a risk intelligent mind-set. In the following chapter, we examine another leading downstream player in this supply chain, also a Sabancı Group corporation, namely, Brisa. But first, an overview of the tire reinforcement materials industry, its major players, and Kordsa Global's position in this global market will set the stage before we go into the details of Kordsa Global's risk intelligent supply chain. To this end, we will discuss topics such as global supply chain risk perception, strategic risks along the supply chain network, Kordsa Global's approach for risk assessment and hedging, global supplier management, strategies for financial and operational hedging including insurance and business continuity, supply chain contract risk management, managing black swans in Swan Lake, and developing talent for the global supply chain mind-set.

In tire reinforcement materials as well as the mechanical rubber markets, the major products are the nylon and polyester yarn, cord fabric, single-end cord, and industrial fabric products.[2] The main tire yarn and fabric producers around the globe are as follows:[3]

- Century Enka Limited (India)
- CORDENKA GmbH (Germany)
- Formosa Taffeta Co. Ltd. (Taiwan)
- Hyosung Corporation (South Korea)
- Junma Tyre Cord Company Limited (China)
- Kolon Industries, Inc. (South Korea)
- Kordarna a.s. (CzechRepublic)
- Kordsa Global A.Ş. (Turkey)
- NV Bekaert SA (Belgium)
- SRF Limited (India)
- Teijin Fibers Limited (Japan)
- Xingda International Holdings Limited (China)

In this global market, Kordsa Global is the market leader for nylon 66 with a hefty 35% market share at the end of 2010.[4] Its nearest competitor holds 30% of the market share. In the polyester market, Kordsa Global is ranked third with a 9% market share. The company has manufacturing facilities in nine countries: Argentina, Brazil, China, Egypt, Germany, Indonesia, Thailand, Turkey, and the United States.[5] Kordsa Global's headquarters is located in İstanbul with the main production facility in İzmit, Turkey. A quick look at the ownership structure of the company indicates that 91.11% of the company belongs to Hacı Ömer Sabancı Holding A.Ş., with the remaining 8.89% publicly traded on the İstanbul Stock Exchange.[6] The market capitalization of Kordsa Global is 815.08 million TL as of April 30, 2012.[7] In 2010, the firm recorded 1,264,097,091 TL in revenue. The European sales, which amounted to 476,598,874 TL in 2010, form the largest share of total revenues.[8] In its transnational supply chain network, Kordsa Global has the following subsidiaries with its percentage ownership structure:[9]

- PT Indo Kordsa Polyester (Indonesia): 100%
- Kordsa Argentina S. A. (Argentina): 100%
- Interkordsa GbRmH (Germany): 100%
- Kordsa GmbH (Germany): 100%
- Kordsa Inc. (United States): 100%
- PT Indo Kordsa Tbk (Indonesia): 60.21%
- Interkordsa GmbH (Germany): 100%
- Kordsa Qingdao Nylon Enterprise (China): 99.5%
- Kordsa Brasil S. A. (Brazil): 94.01%
- Thai Indo Kordsa Co. Ltd. (Thailand): 64.19%
- Nile Kordsa Company SAE (Egypt): 51%

Kordsa Global's leading position in the global market rests on four pillars: (1) high-quality products; (2) ability to provide products on a global scale; (3) product diversity; and (4) superior services.[10]

8.2 Managing Global Supply Chain Operations and Risks

After this prelude, we are ready to delve into lively conversations with Ms. Arzu Öngün Ergene, Kordsa Global's global head of procurement, and Mr. Bülent Bozdoğan, former chief financial officer and global contract manager, regarding the ins and outs of the orchestration of a global supply chain network. In these conversations, first names are used for the sake of brevity. To this end, "Çağrı" refers to the author Çağrı Haksöz; "Arzu" to Arzu Öngün Ergene; and "Bülent" to "Bülent Bozdoğan."

Çağrı: Let me begin with the question of how Kordsa[11] performs its global procurement. How is the global supply chain network organized?

Arzu: Purchased materials are used at our ten global locations in nine countries (Turkey, Germany, Egypt, USA, Brazil, Argentina, China, Thailand, and Indonesia).[12] Consumables and spare parts used at these sites form the integral part of the process. Yet, in our scope, we also still manage the raw materials for yarn and fabric, which are chemicals—and these are all oil derivatives. We additionally manage capital expenditure items and the packaging materials used at each site. These are big-ticket items for us as raw material costs more than 50% of the total costs of our products.

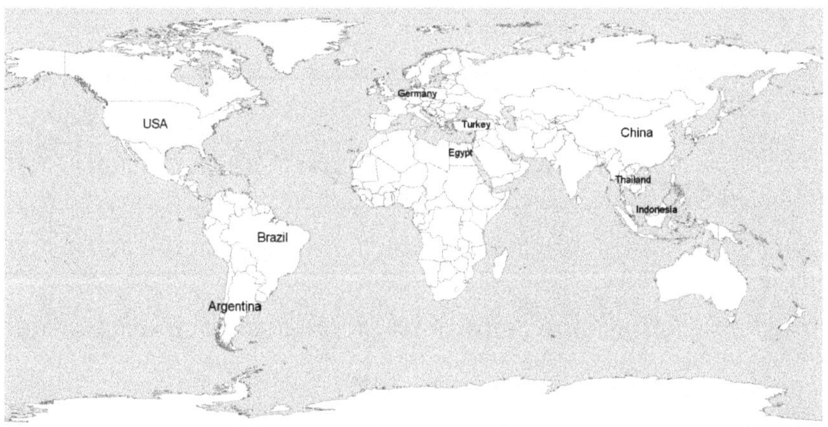

Figure 8.1 Transnational supply chain network of Kordsa Global.

As late Sakıp Sabancı, our former chairman, stated, "Profit is made while buying, not while selling." We therefore try to work under the best possible conditions. But, there is a twist in our process. The volume we purchase does not correspond to the total volume we produce. Thus, we cannot affect market price. We are in most cases price takers in the market. The price formation can be easily understood by focusing on the issues of upstream and downstream along the supply chain. The upstream begins with oil exploration and the downstream sells us the variety of chemicals. Dynamics of such chemicals are within the realm of the oil market. When budgeting for this year, we assumed an oil price within the range of US$80–90, averaging around US$88/barrel, yet now we are seeing US$125/barrel prices.[13] Therefore, in these markets, we cannot control the risks, yet we must be ready as often as possible for any unexpected changes.

The chemical industry is very peculiar. All these chemicals are oil derivatives but upon analysis, use is diversified. Also, substitutes for supply and demand matter. For example, this year the price of cotton has drastically risen. (See Figure 8.2 for the cotton spot price evolution.) Therefore, the price of the substitute material (i.e., polyester) has increased. In fact, the price of polyester has gone up more than 100% in the last year. This is quite unexpected. We could connect this price inflation to the oil prices. However, its price has surpassed the increase in the oil price. So, the supply–demand relationship is of high importance to us.

Furthermore, we are also concerned about supply-related risks. With a few suppliers of the material in the market, our buying power decreases.

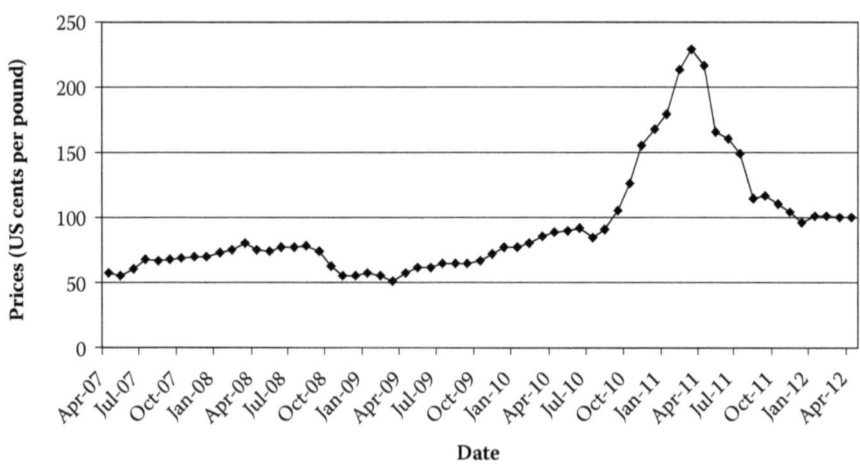

Figure 8.2 Cotton spot market price (CFR Far Eastern ports), April 2007 to April 2012. (From http://www.indexmundi.com.)

Suppliers may decrease their production capacities, shut down facilities, or relocate their assets. Two years ago, one of the producers decided to shut down a large amount of production capacity in the United Kingdom. This movement has made the market tighter. The whole market equilibrium changed in terms of availability and, more importantly, prices.

These actions affect also your customers' future purchasing plans. Thus, these risks are *market-related* ones. Once you go upstream, market risks intertwine with political risks such as how much oil would be produced by the Organization of the Petroleum Exporting Countries (OPEC) as well as conflicts and wars in the Middle East as observed throughout the region.

Çağrı: When you talk about the suppliers, are you referring to the suppliers for nylon, polyester, aramid, rayon, and natural fiber?

Arzu: We purchase polyester chips as raw materials for polyester yarn production, nylon polymer, salt, and two other intermediates (which we use to produce our own salt) for nylon 66 yarn production. We also purchase nylon, polyester, and rayon yarns of several types.

Çağrı: Is the number of suppliers limited for these yarn raw materials?

Arzu: For the nylon 66 part, the number of suppliers is rather limited. Globally, there are four major nylon 66 raw material producers. There are many for the nylon 6. That is the substitute and the competitor product of 66. As mentioned, in the capacity of nylon 66 there has been a "rationalization," which decreased availability. The production technology is a protected one, and capital investment is very high. Thus, a natural barrier to entry exists. The existing suppliers are old and well-established firms. There is only one supplier in Europe, one in China, and two in the United States, and there is a smaller one in Japan. Nowadays, firms move their production bases to Asia.

Çağrı: Have you considered developing a local supplier in Turkey?

Arzu: Sure, for polyester, a local company has recently developed chips that would fit our purposes. If it is commercially feasible, we certainly would prefer to go for local sources.

Çağrı: Which suppliers do you work with at this time?

Arzu: We establish contracts for most of our purchases and work with those suppliers to support the production needs of our entities. We have smaller portions of our purchases in the spot market since it carries more risk. Depending on the market and supplier risks, we may have either *single-sourcing* or *multiple-sourcing strategies*. At this moment, to manage a variety of business disruptions such as *force majeure*, or political events, we employ a multiple-sourcing strategy. However, the more suppliers you have for one product, the tougher it is to handle the procurement effectively. For a certain site, to use more than one supplier, you have to go through various issues such as inventory management and customer approval. We would prefer to have suppliers at different geographies such as one in Europe and one in the United States or Asia. There are tax and customs benefits for such an arrangement as well as proximity considerations. Also, lead times are quite different for those suppliers. Besides, we can mitigate natural disaster risks better in such a portfolio; that is, when a hurricane hits U.S. shores, Europe is safe; likewise, when the Rhein River has water shortage issues in Europe, the American supplier will stay unaffected. Finally, for commercial considerations, to have multiple suppliers for the material is also advantageous for buyers. Especially when the market is tight or down, they compete for business, which creates a healthy market for all of us.

In our industry, which is a complete business-to-business market, we sell the finished products to tire manufacturers, which are very focused on quality and consistency. They would not accept any product before it successfully completes their specific, rigorous, and time-consuming approval process. This process is quite involved.

Çağrı: Because there are health and safety concerns that may lead to accidents?

Arzu: When a car is driven at 180 km/hour, a tire touches the road on an area approximately the size of A4 paper. In essence, you are moving on four A4 papers with that many people in the car.

Çağrı: As you increase the speed, the area touching the road decreases. This may be very risky as we saw in the famous Ford-Firestone Case in early 2001.[14]

Arzu: Sure. This incident has also proven that one of our products called *Cap Ply* is crucial for tires. The nonexistence of *Cap Ply* in that specific tire type was one reason that it burst. The tire surface is inclined to disintegrate as the tire revolves. To prevent this disintegration, not only a belt on the outer surface but also an upper textile called *Cap Ply* is used. If you don't use it, the tires may fall to pieces.

Çağrı: Financial losses as well as fatalities can be catastrophic. Let us go on with the raw material supplier markets.

Arzu: When we talk about polyester raw material producers, it is a more segmented market. There are numerous polyester chip producers all over the globe, but they drive investment in Asia. Hence, the market behaves more like a commodity. On the other hand, this year the market got crazy due to the high price and scarcity of such substitutes as cotton. The cotton prices hit the highest level since the American Civil War.[15]

Çağrı: How do you interpret the reasons for such price movements?

Arzu: One of the major reasons is speculation. Such price movements, especially for natural commodities, are more frequent since in such markets both *seasonality* and *climate effects* exist. Firms ask, "This year we have low yields, so what can we do?" The same issue troubles tire producers as well.[16] They use natural rubber whose price has gone up dramatically. Thus, there are speculative movements, demand growth, growth in world population, and preferences to use some agricultural commodities to others. Sometimes, soy or corn is planted instead of cotton or major producers lower their production. There are subtle interactions that cannot be foreseen on the globe. The production of other crops, for example, affects cotton production, which in turn impacts the price of substitutes.

Let me give an example related to *political* and *regulatory risks* in our supply chain network. China has released a five-year government plan. China is a huge production base for the world and exports enormously. For us, it is a critical supply location. In China's five-year plan, it took measures to vitalize the domestic market demand by reorganizing the taxation system. It is more favorable now for Chinese firms to sell in domestic markets, because its currency appreciated. In contrast, when the Chinese export, until the money flows into the firm, generally after sixty days of transportation, they might incur losses. Turkish exporters were vocal about the higher value of Turkish Lira a few years ago. The same play is on stage in China now. Also, for some products, the Chinese government started applying different rates of VAT return. For example, in some chemicals, only 11% of the value-added tax is returned out of the 17% paid. All of the regulatory decisions mentioned affect our supply chain and procurement decisions. Similar governmental decisions are taken elsewhere; for instance, Indonesia has levied a 5% tax on imported goods.

Çağrı: Is this a new decision?

Arzu: It is new and has been levied retroactively. We have made our budgets, fixed the sales prices, and written the contracts with previous cost assumptions. Suddenly, in mid-February, the Indonesian government stated that beginning from January retroactively, all imported goods would have an additional 5% tax. This figure is extremely high for our industry. We cannot reflect the cost increase to our customers.

Çağrı: Is it due to the profit margins?

Arzu: Sure. This product is not toothpaste. Instead of 1 TL, let me pay 1.05 TL for the toothpaste and we are done! 95% of the customers are *tire manufacturers*. The rest are also industrial firms. Buyers are highly price sensitive. They buy tons of products; thus, the tire industry has very rigid cost and quality management standards. Based on the quality standards, each year we need to improve costs. The sellers and buyers of this industry are both very well trained. It is too difficult to state such a cost increase.

Çağrı: How often is the price being updated in supply chain contracts?

Arzu: In Europe, raw material prices were updated every three months, but in the last year and a half, European firms started to update prices monthly due to a peculiar cycle. In the United States, price updating is monthly. How you build this movement into your purchasing contracts depends on your negotiations. Therefore, it is a great challenge for the industry to manage any mismatch that may arise due to its sales and purchase terms.

Çağrı: When you state the words "the tire industry," you need to examine the whole supply chain. Kordsa Global buys raw materials to produce cord fabric and sell to tire manufacturers. The tire producers sell the tires to auto manufacturers as well as the replacement market. Thus, the consumer buys autos with tires. You have two more stages to reach the "real consumer" in the supply chain. How does the demand volatility in the auto market affect your operations?

Arzu: It affects us dearly. As you mentioned, there are two major customer segments for tires.[17] One is the auto manufacturers; the other is individuals who purchase replacement tires. In the midst of the global economic crisis, when the automotive industry was in a bottleneck, the tire industry did not face such calamities, because even if the customers did not buy new cars, they would replace their tires eventually. They would use the same tires for longer mileage. Thus, when the automotive industry faced a crisis, the tire industry was not directly affected at a similar rate. The replacement market acted partially as a natural buffer.

Çağrı: So, there is a dampening effect in place?

Arzu: Exactly. On the other hand, in the auto industry we have seen interesting dynamics in the last three years. In Asia, the emerging giants such as China, India, and the Middle East created huge demands for automobiles. Especially, in Asia, the demand for two-wheel vehicles shifted toward four-wheel vehicles. Now, with a spare tire in the trunk, you have five tires in one vehicle. This demand has been observed in China and is growing like crazy there as well as in neighboring countries such as Vietnam, Thailand, Indonesia, and Malaysia. Taiwan is in the fast lane already. As the household incomes grow rapidly, auto demand has reached serious figures. Some tire manufacturers have foreseen demand shift and moved their production facilities to Asia in the past ten years. Right now there are many raw material producers in Asia such as natural and synthetic rubber, steel cord, and fabric as well as some other exotic chemicals. Japan is one of the leading countries to house some of these chemicals facilities.

The demand for these materials has shot up. In the last quarter of 2008, when the world was living through a global crisis, when America and Europe were falling down, Asia had also gone down a bit. Yet a dramatically large bounceback has occurred. Nobody was expecting such a fast turnaround. When everyone was discussing whether it would be a W- or V-type recession, in Asia, with a serious bounceback, prices have soared at an incredible speed. For such a high demand, the market reacted with much higher prices, at times artificially created.

Çağrı: I am sure you might have some capacity problems during that period.

Arzu: In such an industry-scale production, decisions have delayed effects. You cannot turn off the oven's knob overnight since there is material in the pipelines or similarly cannot get back to full capacity in one day. But at such uncertain times, you should have your plans to go with restricted capacity usages, even full shutdowns. The opposite is also possible: express deliveries, maximum-capacity usages, overtime on weekends and holidays.

Çağrı: Which locations did you have the capacity reduction in your global supply chain network?

Arzu: You have to follow the market, so if the demand in that geography is reduced, you have to follow. You also have to consider physical capacities of your warehouses. Once the demand is full swing, you cannot easily gain the cuts immediately. You also need to watch your commitments in your supply chain contracts. Parties have commercial and legal liabilities toward each other about volumes and payments. The idea is not to create lost production volumes and thus sales to the other party while keeping your inventory levels as per your own needs. The contracts are for bad days. The good days do not need contracts!

8.3 Perception of Global Supply Chain Risks

Çağrı: What is your perception in terms of the impacts of these supply chain risks? Which of them are the most critical ones in your global supply chain network?

Arzu: If any of these supply chain risks are realized, you are adversely affected. This is similar to what happens to your body if your blood flow is disrupted in any vein of your brain. If it is disrupted, that means the end. You become paralyzed. Your facilities cannot function anymore. Surely, the *market-related supply chain risks* have higher probability of occurrence, yet any one of the supply chain risks will have a significant impact should it occur.

Çağrı: How do you perceive *demand-* and *market-related risks?*

Arzu: I perceive the market-related risk having higher probability than the demand-related risks. Both have high impacts. Most of our facilities work 24 hours a day, 365 days a year. Labor has to be paid. Our energy bills have to be paid. Likewise, if the production line stops in a continuous polymerization line, the polymer flowing through the lines is wasted. You need to shut down, disassemble, clean your polymer lines, and then reassemble them. This is a major financial loss due to not only the wasted polymer in the system but also the maintenance activity and the lost production time.

Çağrı: How do you perceive the risks in the suppliers?

Arzu: The probability in the suppliers would be lower than the market risks, yet it would be closely correlated with the market.

Çağrı: How about *demand-related risks*?

Arzu: We are better at anticipating demand disruptions. As mentioned, as we experienced in 2008, unless we have a much larger crisis, demand risks are manageable for us.

Çağrı: On the demand side, you have longer term contracts.

Arzu: Not necessarily all long-term contracts but long-term relationships that are built on reliability and trust. Also, our customer base is much larger than the supplier base. We sell to all of the major tire manufacturers.

Çağrı: In that case, you could also cross hedge risks of different customers in your portfolio.

Arzu: Exactly. The *customer portfolio risk* is being managed accordingly.

Çağrı: How do you perceive the probability and the impact of demand risks in your supply chain?

Arzu: It would have a mediocre probability with a mediocre impact with respect to others we have discussed.

Çağrı: And how do perceive the *regulatory risks?*

Arzu: The impact is high; we cannot control it. The probability is relatively low. When you discuss political and regulatory risks, you see that they are related. The only issue that would differentiate them could be the risk of war, which is essentially a very low-probability/high-impact event (*black swan*). On the environmental side, we have experienced something that was not envisioned: the Japanese Fukushima disaster. Japan is a major crude oil processor. It processes almost one-fifth of the Asian crude oil. After the disaster, because of its inability to process crude oil, the price of crude oil went down. Yet the oil derivative prices soared dramatically. This kind of environmental and external risk is unlikely to happen often, yet when it happens, it happens.

Çağrı: I am sure you have been through other major *environmental disasters* in your global supply chain. For example, what is your supply chain strategy to mitigate risks due to hurricanes in the Mexican Gulf?

Arzu: This is one of the major risks that might have a negative effect on our global supply chain operations. Our raw materials for nylon are generally shipped from that region. When the hurricane season gets closer, we become anxious. We do always have a *backup plan*. Our suppliers have backup inventories at other locations, and we have backup stocks at the production facilities. If such an event occurs, hands up: *force majeure*! Insurance policies cover such calamities, yet when the business is interrupted, you need to ensure that the incident does not cause the erosion of your customer base. The credibility built with customers and the promises made should not be erased with a single hurricane.

Çağrı: Is there any such recent event you actually experienced?

Arzu: Of course there is. A major U.S. supplier was badly flooded during Hurricane Katrina,[18] and thus the firm cut a major level of chemical production in the market. They were shut down for two years, and they just recently resumed production.

Çağrı: Two years sounds very long to me.

Arzu: I am not an expert to say how much time is needed to recover, but yes it was two years. However, taking this volume off the market had an increasing impact on prices. We are talking about very large capacities.

Çağrı: In your global supply chain network, you are also exposed to *intentional (man-made) risks* such as *terrorist attacks, sabotage,* or *strikes.* These are intentional supply chain disruptions as opposed to unintentional environmental disruptions. Are there any interesting examples you have experienced?

Arzu: These types of supply chain risks are mostly experienced by our supplies in Europe. We are battered by the strikes in France, especially the ones at the ports happening weekly. If such strikes hit the port operations, we purchase over land routes as a backup option. If they do not load the goods, we purchase goods as FOB [free on board] and load them ourselves. However, none of our global production facilities has seen a severe strike until now. Kordsa is one of those firms that manages the workplace peace effectively. It is a fair workplace. If you examine the seniority levels at Kordsa—not only in Turkish facilities but also at global sites—you would see that these seniority levels (both blue- and white-collar employees) are very high. If the strikes hit the other parties' facilities, our insurance policy protects us to some extent.

Çağrı: This is interesting. In the last few years, insurance firms have been working on pricing a variety of corporate risks. Yet not all of them are easily and correctly priced in the market.

Arzu: You are right. You could have everything insured, yet the premium to be paid might not be favorable at all.

Çağrı: There are unknown and unknowable risks that cannot be even thought about. For instance, the Japanese earthquake and the tsunami are not something unexpected. However, the Fukushima disaster after these two events was really unexpected. These are interdependent and cascading risks. How would you price such a multitude of risks? If you could price it, that is perfect. But will anyone actually purchase the insurance at that price?

Arzu: Exactly. The insurance premium would be exorbitant. Moreover, in some countries, the insurance market structure is different. Not all types of corporate risks can be insured. You cannot insure a specific risk in that country within reasonable premium limits. An example is *third-party liability risk (product liability risk).* This is very important for us due to the nature of the criticality of both car tires and aircraft. As you know, when airplanes land, they experience a load of thousands of tons at a very high speed.

Çağrı: It is great that you can manage such global risks.

Arzu: Enterprise risk management has gone a long way at the Sabancı Group. We have an umbrella policy under which major risks are covered. Apart from this, for each country we have a specialized policy that needs to cover *idiosyncratic country risks.* Thus, we can manage some of the *supply-, market-,* and *production-related risks* with these policies.

Çağrı: I understand. Analyses must be conducted for each country. The suppliers that serve certain facilities must be assessed.

Arzu: We are working with global suppliers that serve facilities in Turkey, Brazil, or Indonesia. Based on the raw materials, we have ten to twelve global suppliers that serve multiple facilities in our transnational supply chain network.

8.4 Strategic Risks along the Supply Chain Network

Çağrı: Before we discuss the strategic risks, could you shed some light on the major competitors of Kordsa Global?

Arzu: Kordsa Global is number one in the nylon production and the nylon 66 industrial filament having a noncaptive capacity. Our major competitor is in Asia. Apart from that, there are a number of small producers around the globe. Since European Union (EU) costs are quite high, we cannot say that competition from Europe is severe. Around the world, there are at most ten firms worth considering as competitors. Kordsa has almost 36% of the market share in nylon 66. If you look at nylon industrial filament, our market share is much higher.

Çağrı: Let us now discuss strategic risks along your global supply chain network. These risks refer to technological changes and innovations, new product introductions, and entry of new competitors into the market. They are intimately connected to the corporate strategy of the firm. How have you been managing these types of risks?

Arzu: The pneumatic tire was invented in 1900. Technology has been improving since then. We do have a market intelligence department at Kordsa. The market development and research and development (R&D) groups closely monitor the market, future materials, and world trends. Our R&D team works closely with TÜBİTAK [the Science and Technological Research Council of Turkey], but nothing happens overnight. Unfortunately, with regard to these technologies, Turkey right now is a fast follower, not a developer. In the United States, NASA provides the market with materials that are not used in space projects. In Japan, there are many small-scale high-tech labs. Our colleagues are intimately watching these technological changes with a peripheral vision all over the globe. We need to admit that in today's conditions, to develop the depth of knowledge and the required technology, you have to patiently invest for long periods. It may be in our eighteenth year that there will be some novelty. With the current market mindset, it is very rare to find such highly devoted firms in Turkey.

Çağrı: Since Kordsa has ten facilities in nine different countries, it should have a global perspective in R&D even though your R&D center is located in Turkey.

Arzu: Our colleagues are constantly on the move around the world. They visit each and every customer as well as the supplier. I do meet with suppliers for procurement planning. They do meet for new products and future development. They do not just sit and work in their labs.

Çağrı: How do you perceive these strategic risks in terms of probability and impact?

Arzu: They are low-probability/high-impact types of risks, that is, black swans. Let me give a few examples that will support my perception. First, Michelin introduced something innovative called *Tweel Tire*. It did not contain textile in it, yet it turned out to be expensive and did not fly in the market. It was expected to shake the market, but it did not. It was a tire with no air and had no chance to go flat.

Second, in the last ten to twelve years, we have seen a tire called *Run Flat Tire*. The name may be different for each brand, but the idea is the same. Even if it goes flat, you can still drive the car for another eighty kilometers at a certain maximum speed. It has a special air pressure device inside, and it is used just on high-end autos. So what is the strategic impact? It eliminates the need for the fifth (spare) tire in car, makes the car lighter, and thus helps gas consumption as well. If the cost of this technology can be decreased to take it to all types of cars and not just high-end ones, it would be a game changer and thus a *positive black swan*.

Third, rayon has been replaced first with nylon, then nylon 6, then a better-generation nylon 66, and finally with a low-cost polyester. But did the previous materials disappear in the market? No way! The rayon share stayed at 2% in the global tire fabric production. The nylon 6 and nylon 66 shares change based on their respective market prices. Polyester, due to its price, began to steal a bigger market share among these materials.

8.5 Kordsa Global Approach for Risk Assessment and Hedging

Çağrı: Let us discuss the supply chain risk assessment approach of Kordsa Global.

Arzu: It is surely not easy to assess some of these supply chain risks. The market-, supplier-, and demand-related risks are easier to assess scientifically based on facts. In contrast, the regulatory, political, and environmental risks are not easy to quantify. We examine reports by the World Bank and constantly discuss with financial institutions regarding country-specific risks. If a financial institution changes Indonesia's risk score to a better level, it's a good sign. These financial institutions have better access to financial data than we do. On the other hand, we conduct in-depth analyses every two years for supply- and market-related risks. Furthermore, we have a diverse set of intelligence gathered via different sources. We do subscribe to the daily and monthly reports created by the market research firms that monitor the industry. We participate in seminars and conduct supplier audits and visits. We do work in coordination with marketing teams. We share intelligence and discuss issues in sales and operations meetings. We organize management meetings every year where we assess the market. In those meetings, we discuss the impacts of potential investments and competitor moves. If such an investment takes place in one location, that firm may react in such a way, and then the price may tend toward this direction. Thus, we have a close monitoring of investors, suppliers, and customers.

Even if you have these many closely related risk monitoring capabilities, at the end of the day an unexpected stone may hit you on the head such as the Libyan war, the Fukushima disaster, or a hurricane. In Europe last year, we lived through a natural disaster where the water level of the Rhein River dropped to its lowest in the last twenty years. This prevented barges from transporting goods. Any of these events can occur any time. You cannot mitigate all of these potential risks, yet we closely monitor the aforementioned risk trio (*market, supplier, demand related*) using these methods. We publish and share monthly reports with the rest of the firm for these risks.

8.6 Global Supplier Management

Çağrı: You do have *strategic alliances* with your key global suppliers. How do you manage these alliances?

Arzu: We have strategic alliances especially with the nylon suppliers. For the new ones we work for a trial period. We have recently added new suppliers to our portfolio after such a trial period. We form a team of technical and quality management people. In the initial phases of the relationship, we visit and review and then after a certain period of collaboration, we perform an audit. We examine the production, processes, machinery and equipment, and technology deployed. We also closely examine the suppliers of this supplier—going upstream.

Çağrı: Could you explain the formal process each supplier has to go through?

Arzu: We have a *supplier evaluation system* in place. For all our global suppliers, procurement, warehousing, production, and shipping and handling departments give us feedback. We identify the suppliers that underperform. Giving them priority, we conduct a *formal audit.* Each year a maximum of four suppliers goes through this audit. This close monitoring also opens a path for *collaborative development.* Certainly we do have NDAs [nondisclosure agreements] with these suppliers. Their technology is not freely available. When we have a single or sole supplier (e.g., exotic chemicals), it is also risky for us to protect the intellectual property.

Let me give an example. Recently, there was a serious disruption for one of our additive materials. We buy in miniscule quantities, yet it is indispensible in our production process. Although its financial impact in terms of purchasing is small, its impact on the production is detrimental. In the last twenty years, there have been no problems with the material or the supplier.

Çağrı: So will this event disrupt production?

Arzu: If nothing is done, yes. We took action and transferred some of the inventory in between our sites, and we also found another supplier. Next year, we will audit this new supplier. Accordingly, suppliers that are single sources, the ones that underperform, or the ones that are a black box for us need to go through our formal auditing process.

Çağrı: In this context, have you ever used an independent auditing company?

Arzu: No we have not. Once we discussed the issues with a number of organizations, it became clear to us that their know-how was not sufficient. Third-party auditing is not well developed in our industry.

Çağrı: It is too specific and context dependent.

Arzu: It surely is. They have less experience than our internal teams, so we handle it internally. We use an *apprentice system* and train young colleagues. They are first trained to get their ISO certifications. Then they get their on-the-job training experience working within our teams doing internal audits at our plants. Finally they participate in *supplier audits* along with their experienced colleagues. Besides auditing, we do have a *supplier evaluation tool.*

8.7 Financial Hedging, Insurance, and Business Continuity

Çağrı: I want to ask whether you use financial derivatives such as forwards and futures, options, and swaps to hedge supply chain risks.

Arzu: Our materials do not have open trading markets and exchanges. On the other hand, you could trade the underlying commodity, yet this is completely a new territory.

Çağrı: There is also the possibility of cross-hedging, such as in the cotton markets.

Arzu: You can trade the previous level in the upstream and go for aromatics. Yet you would need a serious team of traders for such operations that need to cover the world markets. Such trading is not necessary at our scale and scope.

Çağrı: Do you use *business continuity insurance* as a risk management instrument?

Arzu: Yes, we do. Business continuity insurance covers the *business interruption risk.* It covers anything that is beyond our negligence. If you forget to hit the button for the machine to start, that is certainly not covered. On the other hand, in August 1999, a major earthquake that was 7.6 on the Richter scale hit the İstanbul–İzmit region with a loss of more than 17,000 people. We got covered for damages and the interruption to business.[19]

8.8 Operational Hedging Strategies

Çağrı: Let's discuss your *operational hedging* strategies. You've mentioned that you could transfer inventory across locations or hold buffer inventory when needed. Are there any other operational hedging strategies you employ at Kordsa?

Arzu: In the last years, we started to deploy global contracts that would encompass the total consumption volume rather than specifying per-site volumes.

Çağrı: Great. You are essentially using an *inventory pooling strategy.*

Arzu: We also have a close watch over the facilities with our sales and operations planning team. Each month, we examine what the facilities committed to purchase or use and the actual realizations. I need to remark that our lead times via ocean freight are long. From the Far East, it takes on average forty to forty-five days. From the United States, around thirty to thirty-five days. None of our materials are being bought next door. Our only facility having such a luxury is located in the United States. They purchase most of the materials directly from the United States.

Çağrı: How do you manage *lead-time variability*?

Arzu: We closely monitor shipments. Even if it is a C&F (Cost and Freight) delivery, we monitor vessel movements daily. If we sense an unreliable situation of the supplier or the chosen carrier, we immediately convert the transportation mode to FOB to gain more control. Moreover, we do hold *buffer inventory* to hedge against the lead-time uncertainty. We do not work in a just-in-time setting. Yet we are keen on effective inventory management. Although extremely costly, in rare cases where necessary, we also use air freight. We employ this transportation strategy against problems downstream in our customers' businesses. In the eyes of our customers, this is what makes Kordsa Global a truly long-term reliable supplier.

Çağrı: Let me also ask whether you have a *global transshipment strategy* among different facilities.

Arzu: We sure do have. We have a dedicated director for such a purpose—the head of global sales and operations planning. This team reviews the dynamics of the system weekly. Moreover, we are in the process of implementing an optimizer software; thus, we can evaluate all the variables and plan ahead for the previous stages in our operations.

8.9 Intelligent Management of Supply Chain Contracts

Çağrı: Could you elaborate on your supply chain contracts and how you manage them?

Arzu: In general, all suppliers prefer that customers purchase in a *linear fashion*. What does it mean? It simply means that the customer buys a certain amount of raw material every month. At the end of the year, the contracted amount needs to be purchased. The more flexibility there is between the periods within the year, the easier it is to manage fluctuations. Hitting a yearly target is easier than hitting a quarterly and a yearly target at the same time. We also let our suppliers know ahead of time, the periods when we will buy lesser volumes such as planned maintenance shutdowns. We also inform the suppliers about our intention to produce so much before that particular period. As we try to incorporate these flexibilities into the contract, I need to admit that we learn by doing and also learn from our failures. There is a Turkish proverb stating, "One calamity is more valuable than a thousand suggestions." Thus, we learn as we are hit by events that never existed before.

Çağrı: Do you generally use supply chain contracts that last one year?

Arzu: They are at least one year long. They can last up to three years. In the good old days, a few contracts lasted ten years.

Çağrı: For these supply chain contracts, how is the price determined?

Arzu: We use a pricing formula.

Çağrı: In one year, the total procurement volume is fixed. How about flexibility on the total volume?

Arzu: We make our budgets based on the annual sales forecasts. In that process, we closely monitor how much we will buy and sell and to which direction a trend grows as well as for which customer. The confluence of this information affects our total production volumes. Essentially, this figure is affected by two major drivers. The first one is *how demand changes,* and the second one is our *production output.*

Çağrı: How much of the suppliers' capacity is devoted to your procurement is also a critical issue. Are there targets you aim for?

Arzu: Our purchase is not as large as you would expect in their total production volume. We are surely in the top ten customers in terms of size. Yet a supplier may prefer to sell to ten different customers on ten different occasions instead of selling to us once. This decision depends on its target market, price expectation, and choice between selling to spot and contract markets. If the price is better for the supplier, that will be its preference. If the spot price is favorable in the market, selling to the market and not Kordsa may not upset the supplier. On the other hand, when the spot market price is not favorable, then any fluctuation in the selling price may be costly for the supplier.

Çağrı: When you compare how much production capacity is devoted to you and your competitors, do you see any major discrepancy that might create a competitive disadvantage for Kordsa?

Arzu: It is more or less the same. The raw material we procure is also bought by the carpet and engineering plastics industries. Profit margins are different for each different market segment. We may not necessarily be the most profitable segment, but we are a relatively safer harbor in terms of size and volatility when compared with some other end uses.

8.10 Expecting the Unexpected: Black Swans in Swan Lake

Çağrı: Now, let us move on to low-probability/high-impact events, that is, black swans. Could you explain the Kordsa Global approach for managing these types of events?

Arzu: We generally conduct scenario analysis. Kordsa focuses mainly on three scenarios: *positive likely, negative likely,* and *most probable.* We also conduct scenario-planning exercises. In such studies, we could examine what events will happen in a bipolar world. In these studies, we also analyze procurement options. In parallel, we consider the *political instabilities* and the *environmental concerns*—especially on the tire side. The EU ratifies new legislation, such as the branding of tires, how they should be disposed, and what kinds of materials they could contain. Based on those scenarios, we work on contingency plans. We cannot incorporate such a clause in our contracts anticipating Japanese earthquakes. Yet we could add some general clauses in our contracts. Suppose the world economy hits the bottom; what will the parties do? Maybe incorporate a hardship clause and renegotiate the contract terms. The question is how parties could incorporate such clauses in their supply chain contracts. It will depend on the supplier–buyer power and the market situation. A contract is give and take. This year, one party may walk away with better terms, but next period it could be a different result. Over time, the terms and conditions should prove to be fair for the parties for the relationship to be sustainable.

Çağrı: What's being considered for the renegotiation process?[20]

Arzu: Mainly, procurement quantity, price, and term.

Çağrı: Have you ever experienced a case where you had to go to arbitration?

Arzu: No, not yet.

Çağrı: I believe such a case is very undesirable for both parties.

Arzu: No benefit accrues to either party in such a case. At that point, when I stop purchasing materials, my production will be disrupted. These clauses in such contracts are beneficial rather than destructive for setting boundaries for both parties. Hence, if those conditions are invoked, both parties will be worse off in the end. A Turkish proverb—*Do not beat the vine grower, but eat the grapes*—nicely encapsulates what we are aiming at in constructive partnerships.

8.11 Future Global Challenges and Emerging Risks

Çağrı: In the next few years, what are the expected events, opportunities, and threats on your radar?

Arzu: For us, a serious concern is in which direction the world economy is going. Kordsa Global is a geographically widespread company across the globe. We are on each continent, except for Australia and Antarctica. This position is critical for us since we require a well-managed assets base with a reliable sourcing strategy. If the developed economies do not grow at meaningful rates, our suppliers in such locations (the European Union and the United States) may discontinue investing in their assets there, maybe license their technology, or move their assets. Overinvesting in the Asia-Pacific region looks like it will also be a headache for the world. These are significant risks on my radar.

Taxes and *antidumping duties* levied in different regions are critical for us. Why? Because even if you source a material from another location at a reasonable price and bring it into a protected economy, such as Brazil, you will face extremely high import taxes. Therefore, it is smarter to have local suppliers in such locations even if the base price costs you more than do the rest of the others worldwide.

Çağrı: This point brings us to the topic of *currency risks* while managing your global supply chain network.

Arzu: Currency risk is relatively easier for us to manage. At some locations we buy and sell in foreign exchange. The problem is at locations where you buy foreign exchange and sell local or vice versa. In both scenarios, there is a pitfall; either your costs increase and sales price decreases, or your competitors with a local currency structure become competitive. Therefore, for us, the locations of production and consumption are very critical.

In Europe, many producers have abandoned the market or have consolidated. Based on the technological trends, these firms are detrending themselves and are changing their production locations. They also change hands. I personally think that Europe will not be "the supply base" in the long run.

On the other hand, to which direction the world will grow is of concern to us. As it seems now, it will grow in Asia. Then, will the rest completely disappear or, due to protective reasons, will they design new rules and regulations? Perhaps the EU legislation will impose higher anti-dumping duties. Or the United States will state that it will go with an inflationist economy and become the largest exporter of the world. Will the United States then swap roles with China? Previously, the United States was sourcing from China; will China now source from the United States? Small events are popping up each day. Sometimes one should move outside of the daily hectic life, zoom out, and examine the events much more deeply.[21]

When you visit China, you get dizzy because of the gargantuan scale, size, and speed. The Chinese economy grows and adds capacity at phenomenal rates. For instance, a firm hits a seventy-five kiloton capacity from twenty-five kilotons in a short while. You visit a production facility. It is five kilometers from one end to the other, a totally vertically integrated process. The crude oil enters as the raw material at one end; the fabric leaves as the finished good on the other end. Hence, the size of production, consumption, capital investments, and state incentives are very different in that region.

Çağrı: We also know that Chinese firms go to the original sources for procurement such as buying African mines for metals.

Arzu: Certainly. For example, they connected with Nigeria and South Africa via SABIP [the Strategic Advisory Board for Intellectual Property Policy]. When I began working at Kordsa, I was on the marketing team. Then, you would see many Koreans at Iranian airports; now, you see many Chinese. Surely, South Korea, as a country of limited resources, creates value via trade and processing. At this point, China emulates what Korea did somewhat relentlessly. I think Vietnam and Thailand have to be closely watched in that region, too.

Çağrı: How about Indonesia?

Arzu: Its job is much harder. It is an island nation. It does have infrastructure as well as human capital problems. In contrast, most probably because of the communist regime, in China the infrastructure is already in place. The transportation network works efficiently. You could also find qualified human capital with some effort and higher cost. For us, as I mentioned, how the world will grow is of critical concern. This will eventually determine our decisions on how we should source, transport, produce, and sell.

Çağrı: At this point, we are living through a change that will continue for the next five to ten years.

Arzu: Actually we live it every day with China.

Çağrı: That is right. Yet I foresee that the system will slowly come to a critical threshold, and then suddenly an abrupt change will occur. This happens in complex dynamical systems with phase transitions. We all know the frog story—of the frog that cannot escape and dies in slowly heated water. The frog is happy in the lukewarm water until it boils, and then there is no time for the frog to escape when the water boils. For instance, in energy technology, such as solar panels and wind turbines, Chinese firms are also quite fast becoming world players.

8.12 Developing Talent with a Global Supply Chain Mind-Set

Çağrı: Now let me move on to a critical topic, which is *talent management.* Talent drives your culture and affects the way of doing business. As a global company, developing, retaining, and managing global talent should be challenging for you.

Arzu: What you refer to as the talent sounds like something related to innate characteristics. In my view, instead of phrasing it as talent, we should call it *sound effort, hard work,* and *discipline.* It is imperative that one covers a wide range of areas to understand interactions. You need to know the intricate details of your market, the rules and regulations, and the language of law. Also, one needs healthy paranoia[22] to more effectively manage contracts and suppliers. The cross-departmental work is also critical. My customer is my company. So the prices we source at should be also desirable for the sales team. Likewise, even if we source at greatly negotiated prices, if the material does not arrive on time and the production is disrupted, it is useless. Also the quality levels should be such that our yields would not depreciate. Therefore, one has to know the company inside out. Personally, the work I did in marketing and finance has helped me now a great deal working in procurement.

Çağrı: Actually it is a great position that amalgamates all of your previous experiences.

Arzu: Not everyone is as lucky as I am. For me, the training I obtained at Kordsa helps me a lot in this position. For example, my experience in finance and the international trade become very beneficial while we negotiate the supply chain contracts.

Çağrı: How about the sales and marketing side of the business?

Arzu: A close collaboration with the marketing team is essential. One needs to be aware of issues in a wider spectrum. For instance, questions such as these should be on your mind. What should my delivery and payment terms be so that I am aligned with sales? How will my decisions affect the inventory levels as well as the balance sheet? In marketing, you get the leading indicators of the markets. Unless one comprehends these indicators on time, you have to follow the events with lagging indicators. There is abundant information and data at this time in our history. You should know how to differentiate what is valuable and what is not in your decision making. We are grateful to the people who founded the Internet. There, the information is linearly available, yet it does not mean the same thing to every one of us. It should be understood and processed for meaning and intelligence.

Çağrı: That requires an adequate and solid background.

Arzu: Sure, that is necessary.

Çağrı: As Rumi said in the thirteenth century, "Not only the answer but also the question stems from knowledge." Voltaire had a similar quote stating that you should judge a man by his questions rather than by his answers.

Arzu: I agree. You need to know the production process and facilities very well. Unless you first see a product in the warehouse, you cannot later understand if the packaging is correct.

Çağrı: Sure. How do you apply these important ideas to your own processes? What qualities do you look for in newcomers?

Arzu: We are seeking analytical people. For example, the analytical approach is necessary to understand how the price evolves over time. Additionally, being able to understand cause–effect relationships is crucial. We also look for general curiosity and boldness in taking initiatives. Because a miniscule price difference of 1 U.S. cent may result in US$1 million with a large volume of procurement. This is a lot of money to give away for any company. You need to be frugal and spend wisely. To be able to provide these qualities to a newcomer is difficult unless there is the right background and upbringing. Last, we look for dedication and a wide vision to see the world and comprehend what you see. You require a somewhat multidisciplinary mind-set.

Çağrı: Individuals should have broad vision and need to work hard to expand it.

Arzu: Also they need curiosity and enthusiasm to expand that vision.

8.13 The Future Vision for Kordsa Global's Risk Intelligent Supply Chain

Çağrı: Do you see Kordsa as successful in the global arena? Is there anything Kordsa could have done better? And what does Kordsa learn from its successes?

Arzu: Kordsa is definitely successful. We have focused on the "Agile Kordsa Global in High Value Businesses for Sustainable Growth" vision in 2012, which personally I think could have been done earlier. Fast and bold moves may help in pursuing some opportunities although risky at times. I think we could have entered and established a stronger base in Asia earlier, for example.

Çağrı: So you mean the big-picture mind-set is lost sometimes?

Arzu: You need to evaluate not only the rate of returns but also the big picture and question your assumptions. Decisions are related to intelligent risk taking, not going off half-cocked. This is my own viewpoint; it does not belong to Kordsa. We do not have another company like Kordsa in Turkey. Kordsa owes it success mainly to our group chairperson, Ms. Güler Sabancı. She took the bold steps when necessary. Thus, it is a unique company per se. On the other hand, there is no other company it can look up to and emulate. We learn by trying to find the best practices that are not necessarily in our industry. Thus, the applicability is not 100%. For Turkey, in my view, playing a complete Western game is a loser. Turkey could have been on a different playing field if we were to choose our own original path like the Asian countries did. We also had internal issues while some nations shot ahead of us. South Korea is a good case in point.

Çağrı: Is there an intention to grow in Asia?

Arzu: If there is any growth, certainly it will be in that region. Kordsa will grow in regions where its customers grow.

Çağrı: Last, what are the most critical success factors in your industry?

Arzu: I would say *integration*. It is the integration mainly at the raw material side of the business. The world is moving toward that direction now. In this way, you both have a steady stream of supply and guarantee a certain level of quality. Moreover, the production processes are under your control. Besides, the major value added is held inside the company.

Çağrı: You mean becoming vertically integrated would be a viable strategy?

Arzu: Sure. Kordsa did it to some extent. We integrated the cord and the fabric. Especially in polyester now, where there is severe competition, upstream producers are entering into the downstream operations, all the way down to fabric, trying to make margins that would keep them alive.

8.14 Managing Risks in Global Supply Chain Contracts

I would like to now offer the lively conversation held with Mr. Bülent Bozdoğan, who has served as the chief financial officer and global contract manager for Kordsa Global during 2001–2009. This conversation particularly sheds light on the nuances of risk management in the context of the supply chain contracts considering the global operations of Kordsa Global.

Çağrı: How do you see your own market vis-à-vis the engineering plastics market?

Bülent: The engineering plastics market is much larger than ours. Also, their prices are much better in that market. Surely the price fluctuates, yet the main customer is the automotive industry. All the plastic components you see in a typical car are made from this polymer, which is also the ingredient of our nylon. Since the price is higher in that market, firms may prefer to use their production capacity for those customers. Sometimes producers may even use global crises and even natural disasters as excuses to curtail production capacity.

For example, the hurricane that hit the Mexican Gulf caused a supplier to declare *force majeure* and not resume production for two years. A natural disaster is not necessary to curtail production capacity. Other facilities were also shut down in North America.

Çağrı: They create product scarcity in a sense.

Bülent: You may call it scarcity or alignment with the market demand. Since the market demand is affected by the same firm, the market price does not depreciate much, staying at a certain level.

Çağrı: It seems that the market is mainly dominated by a single supplier.

Bülent: In the *ADN [Adiponitrile]*, which is the ingredient for nylon 66, there are less than a handful of producers in the world. To establish such a facility, you need at least 100,000 tons of capacity, which means around US$1 to US$1.5 billion investment. This cannot be done by any firm. Moreover, you need to have the know-how and technology. Kordsa surely works on diversifying the suppliers for its raw materials to the possible extent. On the other hand, Kordsa is a secure and reliable customer. These suppliers would like a customer like us even though we seem to be more dependent on them. It is a symbiotic relationship.

Çağrı: Could you summarize the history of Kordsa with DuPont? How did the relationships with these suppliers change after the joint venture with DuPont ended?

Bülent: We had the relationship with DuPont much earlier than working with these suppliers. That is, when Kordsa was the only manufacturing facility in Turkey, there was only fabric being produced. We did not have cord production. We were purchasing the yarn from DuPont. Then we decided to co-produce the yarn and founded a company called DUSA in İzmit, Turkey, next to Kordsa. It was a 50–50 joint venture. DuPont was providing the raw material, and Kordsa was purchasing half of the production whereas the other half was going to DuPont. We were producing the cord fabric, yet DuPont was only in the production of the yarn. There were other converters like us producing cord fabric from the yarn.

When a firm was founded on a global scale, DuPont brought its cord facility where Kordsa put forth the fabric facility and the corresponding know-how. At that point, Kordsa had the manufacturing know-how, standards, and quality level accepted by the tire producers around the globe. DuPont had operations in Chattanooga, Tennessee, as well as in Argentina and Brazil. The Brazilian facility was focusing on the nylon 6 whereas our focus is right now mainly on the nylon 66.

Çağrı: How different are the nylon 6 and the nylon 66?

Bülent: They are made from completely different raw materials. They are both nylon yet have different features. They both could be used in a certain tire. The raw material of the nylon 6 is not being produced by DuPont. It is called *caprolactam*, an oil derivative, and has many more producers around the world.

Çağrı: So it is a competitive market.

Bülent: Sure. It is a commodity product and has multiple suppliers in China, Russia, Europe, as well as other regions. Its price is also more volatile. It increases with oil prices. But the price of anything is now determined by China. If China comes in and purchases, prices go up; if it does not purchase, prices tumble. You should not purchase in the cycle when China purchases if you can. In the tire industry downstream, when China steps in to purchase rubber, the price of rubber shoots up.[23] Once China purchases and stocks and then you enter the market, there is a chance that you might purchase at lower prices. You should not purchase when China purchases if you have sufficient stock levels. It is not easy to manage this process since in the last few years, the behaviors of Chinese firms have deeply affected the markets in a growth environment.

Çağrı: This is so critical. So you need an "intelligence-gathering mechanism" that follows diligently who purchases what, at which volumes, and when.

Bülent: Natural rubber is such a commodity. It is traded in the plantations of the Far East. In China, dynamics are different. Government-aided companies can evolve into major competitors. They integrate backward and establish long-term purchasing contracts from the same supplier base that we use since as government-supported entities the cost of investment and profitability are not real concerns for them. The suppliers of raw materials have also announced plans to invest in China; after all, they have a huge customer base with a strong growth prospect. If these volumes become much larger; we will observe that China will be a major force to determine the market price.

Çağrı: In your industry, what are the most critical success factors?

Bülent: You need to follow closely what is going on in many markets. I mentioned a few critical items. If the Chinese purchase, the prices will be affected. The mode of the global economy, whether it is expanding or contracting, also matters. They all impact the prices for raw materials. When the American economy was in a recession, the automotive sector was in decline, the engineering plastics market was not a critical threat to us. It was not growing, and demand was limited. Thus, if you had contracted in a long-term fashion, they might have protected you when the markets were upside down. Surely, long-term contracts can protect you up to a point. The Turkish proverb "Being tough gives the game away" means there are cost penalties for these contracts if they are breached. One could breach by paying these costs. However, it is essential that both parties respect a contract if they are reliable firms. This is expected, but in the end firms may leave no avenue unexplored to breach because there is such an opportunity. Assume that you have a contract for 1,000 TL and the market hits 3,000 TL. This would create a stomachache for one party. Thus, a supply chain contract cannot last long if there is a positive or negative impact on one party.

Çağrı: Sure. It cannot be sustained for long.

Bülent: The other firm may endure the difficulty if the contract maturity date is near. Yet when the time comes, it will look for ways to get even with you. Or it may suggest that you should improvise a solution. In all contracts, there is a term stating that if the situation is not sustainable for one party, that party can ask for a renegotiation due to hardship. We had such issues in the past and surely will have in the future. Thus, one of the critical success factors is to design and sign a correct contract at the correct time and place.

Another critical success factor is not to cause a production disruption in the facilities because the process does not end with good contracting. The shipments have to be made on time. You need to manage your inventory so that you don't carry too much stock to increase working capital as well as too little inventory to create a stock-out risk. We make the supply chain contracts globally in a centralized manner yet manage the shipments locally. Our colleagues responsible for the global supply chain contracts act as solution providers. Their coordination capabilities are critical to manage timely and correct shipments to global facilities. Surely, the real issue is to establish a close alignment with the sales team at Kordsa.

Çağrı: Sure. The demand side of the supply chain has to be aligned with the supply side, that is, the procurement process.

Bülent: Exactly. It is important to have a mechanism to reflect the cost increases in the procurement to the sales process. To be honest, this is not at all easy. It depends on how close your and your competitors' profitability goals are. You will prefer to continue to sell at favorable profit margins if you have already established such a customer. Assume you are selling at a 20% profit margin, whereas your competitor is fine with a 10% profit margin, if there is a cost increase in the raw materials, you may have to reflect this change directly to the customer. It becomes a trouble to be able to hold the same profit margin if the competitor does not reflect the cost increase. So expectations are driven by the firms' profitability goals. In our industry, customers do not easily switch to other competitors just for the price considerations if they have made long-term contracts. In rare cases, we experience such actions. On the other hand, when you work with long-term contracts, at the end of the contract period, the customers seriously examine the prices. It might be hard to reflect any cost escalations in the prices due to sensitivities of the sales team. Of course, when the world crude oil price jumps to US$150 from US$50, everyone reflects such a cost increase. This is quite understandable.

Çağrı: As you mentioned, on the procurement side, where there are certain *hardships*, the *renegotiation* may take place. In a similar vein, on the sales side, there should be flexibility in the terms.

Bülent: Previously, we had yearly contracts with the large firms. Within the year, you would bear the plus and minus deviations. Nowadays, most contracts are reviewed every three or six months.

Çağrı: On the procurement side, you work with much shorter supply chain contracts, that is, one-month durations.

Bülent: That is right. The durations have been cut. Everyone prefers to reflect the cost on the price. However, still on the sales side, we do not see the monthly contracts.

Çağrı: I think the market shares of the customers should play an important role in this process.

Bülent: We have a double-edged sword. We procure raw materials from major suppliers that are powerful. We sell to the customers that are also major and powerful firms. You are in between these powerful sets of firms. Despite the fact that you are a major player in your industry, you create competitive advantage with the service and quality you provide. In the downstream, they are selling the tires to global OEM giants.

Çağrı: In terms of its position in the global supply chain network, Kordsa Global manages both sides intelligently without creating any losses from either side. This is in fact a great challenge.

Bülent: Exactly. Our management challenge is in fact an art. Automotive firms use enormous amounts of different parts and components in car manufacturing. In that composition, our products go into the tires. The tires are perhaps the most critical part of a car since they touch the ground. In any of these items, the auto makers strive to obtain the lowest possible prices. You should be thankful if you make one extra cent on a single item. Since the volumes are large, it amounts to a meaningful sum. Thus, in the original equipment manufacturing, you do not expect to make a lot of money. Yet you sell at larger volumes since a typical car would have five tires with a spare. In the replacement market, though, you may create more value directly from the end users. However, the profits and the prices would be very different. Both markets have to be managed carefully at the same time.

Çağrı: Let us now discuss your global supply chain network and how it is being managed. You currently manage ten different facilities in nine countries. Globally, there are so many failure examples due to the culture mismatch. You surely have different cultural mind-sets and backgrounds in these countries. It is an interesting challenge. How does Kordsa manage this challenge?

Bülent: We have a technological know-how in the cord fabric production. The school for this know-how is in İzmit, Turkey. When DuPont left the JV, and Kordsa became a 100% Turkish capital-funded company, our colleagues who had been trained in İzmit moved to manage global production. We also revised our management model. Around 2008, we had regional management teams. The regional mind-set has been used since DuPont was part of the game. We still have regions for the global supply chain network. Our first region is Turkey, which is called the Middle East and Europe. The second region is North America. We operate both cord and fabric production facilities there. The third region is South America, and our fourth region is Asia. Previously, we had a management committee composed of regional presidents, with the title vice president. They managed the regions as well as acted as the vice chief executive officer. Besides four such presidents, there was the chief financial officer—myself at that time—and of course the chief executive officer. Now, we've eliminated the title and position of regional presidency and recreated it as site manager. We still have global and regional sales managers.

In our production facilities, we decided that the focus should be on technology and high-quality, low-cost production of the products demanded in the market. Each facility is a legal entity within an integrated firm. Finance continued to be a global function. We formed a corporate finance team at İstanbul headquarters [HQ], yet each facility manages its own finances. All facilities report directly to the HQ, and the consolidation is made in İstanbul. We act as the money management or the funding center since we are geographically scattered around a global terrain. For example, in Europe we have no problems with fund management. The center is in the Netherlands, with only a one-hour difference between Turkey and continental Europe. At the end of the day, all the money that flows to the Netherlands allows us to manage the funds effectively. On the other hand, you cannot do that so easily in America, for a number of reasons. You could not transfer your funds from Argentina. If you desire to transfer funds from Turkey, there is a seven-hour time difference. Later, you could not get back to Turkey due to the inappropriate legislation. In contrast, Europe is a union, and thus an integrated legislation exists. Asia is a completely different story with Indonesia and Thailand.

We manage the different sites by having local managers at different levels from those cultures. Yet the top manager for the site is a colleague trained in İstanbul. It is difficult to know the nuances of the culture and how it affects work and management style. As you know, there is the whole spectrum of management styles from American to Japanese modes. You create a by-law in Turkey. Then you see that it does not apply to a foreign site. Thus, you need a high-caliber team and diverse set of individuals to manage such a global supply chain network.

Let me give you an example. In Asia, the culture is very sensitive when you communicate in a *loud tonality.* On the other hand, in Latin America, not only loudness but also *fun* and *commotion* is widely acceptable. It is very different.

During my tenure, I have organized global finance meetings for seven years. During the day, we had intense discussions, but later we socialized during the evening. Communicating face to face is essential. In a global supply chain network, one facility ships the cord to Indonesia; the other one ships the fabric to another location. These people talk to each other only remotely and with whom in reality? We created a much-needed social cohesion. Besides finance meetings, which were more functional, Kordsa Global also holds frequent meetings for the *Global Leadership Team.*

8.15 The I-Quartet Model of Kordsa Global

Having discussed Kordsa Global's approach and practice to risk intelligent supply chains, we are now ready to wrap up the chapter with a summary of the I-Quartet model roles. The I-Quartet model for Kordsa Global is summarized in Table 8.1. From this table, we can easily see that the risk intelligent supply chain (RISC) of Kordsa Global seems to be very strong on all four I-Quartet roles. I need to note that the Improviser role is generally the most invisible, latent one among the others in the I-Quartet model. Its positive impact is seen in the overall mindfulness and vigilance of Kordsa Global. Besides, the other three roles need to continuously interact with the Improviser role of Kordsa Global's RISC to achieve the most resilient, diverse, and intelligent risk taking supply chain. When one examines the I-Quartet model of Kordsa Global, it will be clear that it is not by chance that Kordsa Global leads in its industry as a pioneer as well as an example emulated by others.

Now we move one stage downstream in the global tire supply chain and focus on Brisa in the next chapter.

Table 8.1 The I-Quartet Model of Kordsa Global

Integrator	*Inquirer*	*Improviser*	*Ingenious*

• Global integrated network of materials, cash, information, and people
• Managing global risks such as "third-party liability" using local solutions
• Integrated intelligence sharing across supply, marketing, sales, and production teams
• Smart management of global purchasing considering taxes and antidumping rules
• Global supply chain contracts integrated with local shipments
	• Market intelligence department to monitor trends in future materials and markets (e.g., carbon fiber, nanomaterials)
	• In-depth analyses every two years for supply and market-related risks
	• Diverse set of intelligence discovered by:
	  • Daily and monthly reports by market research firms
	  • Supplier audits and visits and seminars
	  • Annual market assessment meetings for investor and competitor moves
		• Improvised solutions to manage contracts to preclude breach of contract events
		• Using concessions and trade with other contracts in other regions
		• Intelligent use of renegotiation process to mitigate breach of contract risk
		• Multiple-sourcing strategy[24]
		• Volume flexibility in the negotiated supply contracts using demand and production yield information
			• Supply contracts with abandonment option (e.g., terminating a few contracts in the global economic crisis due to low demand)
			• Managing natural disaster risks such as hurricanes in the Gulf of Mexico with backup options, extra inventory in the European Union and production facilities
			• Managing strike risk by purchasing via port versus land as well as FOB purchasing

(continued)

Table 8.1 The I-Quartet Model of Kordsa Global (continued)

Integrator	*Inquirer*	*Improviser*	*Ingenious*

• Global capacity management (shifting, expanding, and contracting capacity in the global supply chain network)
• Growth via acquisitions and organic means
	• Constant learning from failures
	• Strategy for intelligent learning: "One calamity is more valuable than a thousand suggestions"
		• High-caliber and diverse set of people in terms of culture, ethnicity, and background managing the company's global supply chain network	• Effective operational hedging strategies
			  ◦ Inventory pooling among global facilities achieved with flexible quantity contracts
			  ◦ Transshipment strategy at the raw material level among global facilities
			• Strategy used for intelligent risk taking: "Do not beat the vine grower, but eat the grapes"
			• Sustainable supply chain contracting

Chapter 9

Brisa: Virtuoso of Supply Chain Risk Management

Start your journey in safety; let your journey be safe and sound!

—Motto from Brisa Marketing Campaign

9.1 Prelude

Presenting the case of Kordsa Global in the previous chapter has brought us one stage closer to the final customer in the global tire supply chain network. At this supply chain stage, our case in point is now Brisa, a joint venture company between Bridgestone Corporation and Sabancı Holding A.Ş. A discussion of Brisa's risk intelligent supply chain practices would first benefit from a brief overview of the global tire industry, its major global players, and the position of Brisa in such a competitive market.

Total market value of the global tires and rubber market in 2010 was US$124.7 billion. However, the industry went through a decline in the years of 2008 and 2009, at 3.6% and 6.7%, respectively.[1] The major industry products are the offerings for the passenger car and light truck market segments. The sum of these two segments accounts for more than half, about 59.5%, of the global tires and rubber market. The second largest segment of the industry is the truck category, with a hefty 29.5% market share.[2] The global market share is divided among four industry players as of 2010: Bridgestone (16.07%), Michelin (14.81%), Goodyear (11.15%), and Continental (5.33%). The rest (see Figure 9.1) is shared by Pirelli, Sumitomo Rubber, Cooper, Yokohama, Hankook Tyre, and other firms.[3]

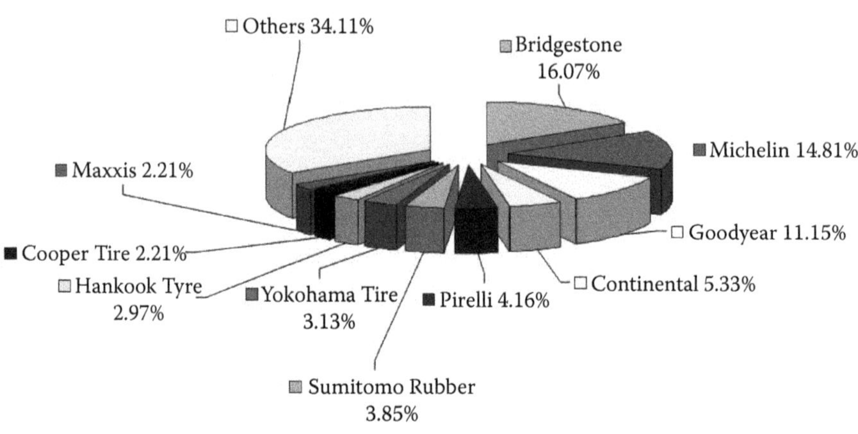

Figure 9.1 Global tire market shares in 2010. (From http://www.bridgestone.eu.)

In terms of production volumes, Bridgestone's total tire production volume was 1.86 million tons in 2008, 1.43 million tons in 2009, and 1.77 million in 2010 with an overseas production rate of 66%, 69%, and 69%, respectively.[4] To lead the market, Bridgestone has developed nonpneumatic tires that are airless and thus need less maintenance. This new production technology additionally eliminates tire punctures.[5]

Brisa is the leader in the Turkish market and the seventh largest tire producer in Europe. In the 1960s, Goodyear and Pirelli were manufacturing pioneers in Turkey. In 1974, Lassa Tire Manufacturing and Trading Inc. was established. A partnership agreement with Bridgestone was signed in 1988, changing the company's name to Brisa Bridgestone Sabancı Tire Manufacturing and Inc. In 1996, Brisa won a number of awards, including the European Quality Award (EFQM), Environment Award (by the İstanbul Chamber of Industry), and Best Performing Supplier Award (Automotive Industrialist Association). Moreover, in 1999, Toyota awarded Brisa the "Top Scoring Supplier Award." This success has continued to today with more awards and honors, including Bridgestone as sole official tire supplier of Formula 1.[6] Today, Brisa exports its products under Bridgestone and Lassa brands to more than fifty countries that are mostly European.[7] As of February 2011, its market capitalization was US$696 million.[8]

Brisa's production facilities are located in İzmit—an industrial city around 110 kilometers from İstanbul—in a 260,000 square meter area. The production capacity was 9,059,051 and 7,406,284 in 2010 and 2009, respectively.[9] Brisa sells its products to original equipment manufacturers (OEMs) and replacement market segments. Brisa serves Toyota, Renault, Ford, Fiat, Honda, Hyundai, Mercedes-Benz, BMC, Mitsubishi, and Man in the OEM market. Besides working with these big names, Brisa's foray into the replacement market segment has generated 8% sales growth when compared with 2009.[10] The Brisa product range includes the following tires:[11]

- Ultra-high-performance tires
- High-performance tires
- Standard car tires
- Winter tires
- 4X4 vehicle tires
- Light duty commercial vehicle tires
- Steel radial van/light truck tires
- Bias-ply van/light truck tires
- Steel radial bus/truck tires
- Bias-ply bus/truck tires
- Agricultural tires
- Off the road tires
- Forklift tires
- Construction machinery and heavy equipment tires
- Motorcycle tires

In 2010, Brisa invested US$28.8 million in capacity expansion and modernization.[12] In the last few years, the company has initiated many successful marketing campaigns, including "Start your journey in safety, let your journey be safe and sound"[13] and "The Safety Tire." One example of the campaign logo for the "Safety Tire" is given in Figure 9.2. These marketing campaigns aimed to increase the awareness of drivers regarding traffic safety and accidents along with suggested precautions.

Figure 9.2 Brisa marketing campaign logo—Safety Tire. (Courtesy of Brisa.)

9.2 Managing a Global Supply Chain Network

Conversations with top supply chain executives of Brisa provide insight as to how Brisa designs and manages its risk intelligent supply chain. To this end, we will discuss topics such as global supply chain network management, supply chain contracts, regulatory and political risks, key risk mitigation strategies, management of black swans in Swan Lake, business and supply continuity management, and supply chain talent recruitment and development.

In the conversations of this chapter, for the sake of brevity, first names are used. To this end, "Çağrı" refers to Çağrı Haksöz; "Fatih" to the supply chain director Fatih Tunçbilek; and "İdil" to the supply chain risk management coordinator and product supply planning manager İdil Z. Taner Ertürk. As Brisa has stated in its marketing campaign, I offer the following motto as we begin: "Let us begin our journey with safety and let the journey be safe and sound as well as exciting and provoking."

Çağrı: How does Brisa organize and manage its global supply chain?

Fatih: In our global supply chain, we have procurement and logistics functions. Logistics deals with all the activities after the products are procured such as the logistics tasks during the import process as well as the coordination with the customs agent firms. Although we work with customs agents, the logistics coordination is done by our team. This team is also an expert in international trade.

Çağrı: Could you elaborate on these activities?

Fatih: Transportation and warehousing of materials are under our responsibility. We also handle the logistics process until the materials go into production. We feed the production schedules, yet that process is managed by a different team. Once the finished goods are ready, warehousing and domestic and foreign shipments are also managed by our logistics teams. We have one central distribution center that is located next to our production facility. We also have small, 100–200 square meter sized depots in our network for some special items in İzmir and Ankara. You cannot call them a distribution center, though. Our daily capacity for outgoing shipment is 1,000 tons. Besides, our product supply planning team consolidates all the information and the sales forecasts from each sales channel, balancing the demand–supply. Then they prepare the master production schedule. Later, this master schedule is executed in daily schedules that are prepared by our industrial engineering team. Each machine works based on the twenty-four-hour daily schedule. We have two procurement teams: one for the imported raw materials and the other for domestic materials. We linked our domestic procurement team to the *OSAT [Collaborative Procurement]* platform. This platform was catering to the firms in the tire industrial group of our parent company [Sabancı Group]. Later, when a few firms abandoned the system, we took over and revitalized it for our purposes. Now this system has 6,000 registered suppliers. We work with approximately 1,500 active suppliers via this system.

Çağrı: Could we call this platform a *private marketplace* that Brisa has developed?

Fatih: To some extent. OSAT does not manage the raw material procurement. It handles the procurement of domestic machinery, subcontractor services, and spare parts. Requests for procurement are obtained; suppliers provide their proposals on the Web. It began operating around 2000. Within years, it became integrated with SAP. Once the approvals are given within this system, each one of the procurement jobs is automatically downloaded onto SAP. Later, SAP updates the information in an integrated fashion. There is a bidding mechanism. Yet it does not work like a regular auction. We ask for bids for a specific tender to all our suppliers on the Web. They provide their bids until a given date. We do a prescreening and elimination and ask for a second round of bids. In this process, suppliers do not know who is also in the process; thus, it is a blind system.

Çağrı: How about the foreign suppliers? How do you work with them?

Fatih: First let me remind you that 85% of our raw materials come from foreign sources. Only 15% is purchased domestically. Therefore, we work closely with Bridgestone Global. We manage the supply chain contracts of major global suppliers together with the Bridgestone Global Purchasing team. We manage the rest of the suppliers—as the Brisa team—located in Europe as well as Russia and Egypt. For those raw materials that are supplied in monopoly markets, we collaborate with Bridgestone to obtain a secure and reliable stream of supply with reasonable prices. Our production capacity in terms of units was 9,059,051 and 7,406,284 in 2010 and 2009, respectively.[14] Globally, we were in thirty-ninth place in terms of revenues.

Çağrı: How is Brisa governed within the Bridgestone network?

Fatih: Brisa is a joint venture firm between Sabancı Holding A.Ş. and Bridgestone Corporation. A total of 43.63% of its shares are equally owned by the two groups. The rest is public, traded at *İMKB* [İstanbul Stock Exchange].[15] It is interesting to note that we are the only joint venture where two partners have equal share in the Bridgestone Group. We conduct meetings with the Bridgestone Global Purchasing team. We discuss the general management strategy as well as supplier and raw material issues in detail. In raw material procurement, you do not have too much flexibility to select your suppliers for our products. It is a special process and requires a set of technical approvals along the way. Thus, if we would like to add a new supplier to our supply chain network, there is a technical approval process that takes one and a half to two years. Sound arguments for such a change need to be prepared diligently.

Çağrı: Could we discuss your major brands in the marketplace?

Fatih: Brisa is the name of our company. We have two major tire brands, Bridgestone[16] and Lassa.[17] Lassa existed before the joint venture. Recently, we have created another brand called *OtoPratik*.[18] Customers use our auto service network until the warranty period ends. Once the warranty period is over, people use mom-and-pop style auto repair services. *OtoPratik* aims to provide a reliable and trusted service for those customers. In addition to tire change, we provide a set of services to our customers such as motor oil replacement, brake system checkup, and air conditioner checkup.

Çağrı: So you aim to have a service chain with *OtoPratik*?

Fatih: Certainly. At this point, we work with the dealers who own these shops. Yet we control the know-how and the management. We tried to create a business model that would enhance the profitability of the dealers. Besides selling and servicing tires, dealers should be able to make money on other services. The profit margins on tires are going down. This brand is therefore created to make Lassa and Bridgestone brands more competitive.

Çağrı: Could we now discuss the market segments you focus on and your specific products?

Fatih: We cater to almost all segments of the market with tires ranging from 12 to 38-inch wheel rims, that is, automobiles, buses, trucks, and construction and agricultural equipment except motorbikes. In terms of stock keeping units (SKUs), we have around 600 products. And our daily product variety is around 180. We produce around 400 items of 600 every month.

When you examine the components of a tire, twenty to twenty-five different components enter the tire assembly process. Thus, it is a very challenging assembly from an SKU planning perspective. When you examine the market segments, there are two major sales channels. One is the *automotive (OEM) segment*, and the other is the *replacement market* for individual customers like us. The OEM segment is an important one. It is the channel that challenges the tire manufacturers because most development requests are initiated from this channel. These requests are in the areas of product development and improvement in terms of weight and rolling resistance. Besides these two main sales channels, we have the export sales channel. This is restricted to our Lassa brand. We sell our Bridgestone tires, which are stamped as *Made in Turkey*, to Bridgestone Europe as an off-take where they are being sold in Europe.

On the replacement market, there are fleet companies that purchase large numbers of tires. There are fleets in the automobile rental business as well as truck, bus and coach, and 3PL companies. We have a specific team that focuses on these fleet customers. Our focus surely is more on the truck and bus fleets. Besides, we have a fourth brand called *Bandag*. In 2007, Bandag was acquired by Bridgestone Corporation. Later in 2011, Brisa acquired Bandag Turkey. Bandag was a global firm founded in 1957 in the business of tire retreading. Retreading is a critical process, especially for the truck market. Once the tire casings are worn out, these parts are cleaned and retreading is done. This process extends the lifetime of the tires.[19] If the tire is scrapped after 300,000 kilometers, with retreading, you can use it for another 200,000 kilometers. This is indeed very attractive for truck users. You can extend the total lifetime of a tire up to 1,000,000 kilometers with two retreading processes. Hence, in managing Bandag operations at Brisa, we aim to coordinate our dealers and the fleets for retreading. We are moving toward being a service provider as well as a tire seller. We can state that when you buy tires from us, you also get the retreading service. Thus, we sell a product that can run for 800,000 kilometers.

Çağrı: Essentially, you are extending the lifetime of customers with Brisa.

Fatih: Exactly. We also visit fleets in their own garages and measure tire pressures and tread depth. We advise on these issues. Due to correct use of tires, you can save 3 to 5% from fuel consumption. This is especially critical for 3PL firms. Generally, 40% of the transportation cost is due to fuel consumption. A total of 3% of the 40% becomes a major cost-cutting item. Besides, when you use the tire at low pressures, its lifetime decreases due to undesirable wear. Thus, we provide on-site advising services and educate our customers. For retreading, timing is critical. It should not be later than the optimal retreading time; otherwise the tire body gets damaged.

Çağrı: In the OEM market segment, who are your customers? How do you manage this channel of the supply chain?

Fatih: We sell to OEMs in Turkey. We supply to each one of the segments, such as passenger cars, trucks, buses, mini buses, and trailers. Thus, our OE customers consist of car manufacturers such as Ford Otosan, Toyota, Honda, Renault, and Tofaş-Fiat as well as truck and bus manufacturers such as Mercedes-Benz, BMC, and MAN. We also work with commercial vehicle producers such as Karsan. On the other hand, our customer portfolio also includes the tractor makers such as Türk Traktör Fabrikası, John Deere, and Erkunt Traktör. Also, some minor local tractor makers purchase tires from our dealers.

All of the sales contracts are designed and executed by Brisa. We could sell Lassa overseas. Besides, we have a limited capacity. The perception that customers have regarding the tires on the car—that tires are original components of the car—is rapidly changing. Customers do not have the following perception anymore. If I have Lassa on the car when I buy, it is better to replace it with Lassa. The market became much more competitive with the introduction of a large variety of brands. You cannot easily assume to get a market share from replacement customers based on the OEM sales. If the customer is very satisfied, there is a chance, yet this is not a precondition right now.

Çağrı: Could you give us an estimate of your capacity allocated to the OEM market?

Fatih: Our strategy is to hit around 20% of the total capacity in this market segment. Since the last economic crisis around 2009, we have been operating in a production bottleneck. Especially toward the end of the crisis, we had seen tax reductions and domestic and European incentives for the automotive market. This inducement has rapidly increased the tire demand. First, it is not a profitable market segment. Second, it hit a point that would damage our other segments. Hence, we decided to intelligently control our capacity for this segment.

Çağrı: How about your competitors?

Fatih: Good question. In Turkey, automotive OEMs are turning toward foreign tire suppliers when Brisa, Goodyear, and Pirelli become short. Pirelli and Goodyear are physically located close to us in this region. Goodyear has two production facilities in Turkey. Pirelli has its own facility as well as a subcontractor that produces certain passenger radial tires. Apart from these two giants, another firm (Petlas) also produces for all market segments. Their market share hit around 3%, yet it aims to increase it with new investments such as tractor rear radial tires.

Çağrı: In the Turkish market, what is the position of Brisa vis-à-vis its competitors?

Fatih: We are the market leader with our two brands. Our vision is clear: We want to be the first and the second in every product segment. There are so many different product segments. If we are the first in one segment with Lassa, Bridgestone should be the second. If Bridgestone is the first, Lassa should be second in that particular segment. Right now, we are for sure either first or second in any segment, yet sometimes we become third or fourth with a set of segments. When you take the sum total of the two brands, our market share is approximately 33%. Then you have Goodyear and Pirelli, each having 15 to 17% market shares. Following them, you have Michelin, Continental, Petlas, and the rest of the small players. That is, 60% of the market is dominated by three companies: Brisa, Goodyear, and Pirelli.

Çağrı: Now let us discuss how you manage the sales side of the supply chain.

Fatih: Sure. There is the customer relationship management team on the sales side, managed by Mr. Timur Akarsu. This team is also under our governance. They deal with the customer orders. We have a state-of-the-art information technology (IT) infrastructure for customer order and sales management. It is the one and only system in Turkey and probably unique for its scale and scope within Europe. Almost all of the sales orders are automatically managed with 1 to 2% manual operations. This team is also our window to the customer world.

Çağrı: When you look at the whole supply chain process, how did you organize different functions?

Fatih: In our supply chain management process, we combined five different functions and aimed to create synergies across them. Product supply planning was managed by the vice president of sales. Procurement was managed by the chief financial officer. Logistics was managed by the business development directorate. OSAT was in a completely different location. Last, the customer relationship team was managed by the marketing group. Since 2010, we have been experiencing a learning process regarding the best model to manage the supply chain coherently.

In March 2011, we decided to use the *Supply Chain SCOR model.*[20] When we analyzed this model in depth, it really fit in well with how we managed our organization. It focuses on all functions apart from the sales and product design. Now we are working on improving our sales forecasting process, which is the heart of supply chain planning. We are moving toward a model where we can compare our statistical forecasting models in terms of financial and strategic objectives. Hence, our product supply planning team integrates the whole process with sales, marketing, and procurement.

The operations go through a series of decisions. Let me elaborate on these. Each month, after we obtain the demand information, we update our three months' demand forecast in a rolling horizon. Once we do this, if we see an issue on the procurement side, we revise our plans. Thus, it is tightly integrated with the procurement process. Right now, as a supply chain team, we are located in the same office area. This setup seems to be very useful. Previously, once changes were made and uploaded to the material requirements planning (MRP) system, other related parties would be informed. That moment would be too late to take action. However, right now, the procurement team members can overhear changes in the sales. We also share our forecast updates in monthly meetings so that we see our direction, potential deviations from our budget, and also potential problems in the supply chain.

9.3 Managing Supply Chain Contracts and Related Risks

Çağrı: How do you manage the raw material procurement?

Fatih: We have long-term relationships with raw material suppliers. We do not purchase in the spot market. We need approved suppliers to work with. Once we have a number of suppliers, we use a *dual sourcing strategy*. That is, we procure, say, 60% from supplier A and 40% from supplier B to manage supplier-related risks. We decide on the percentage allocation on a yearly basis. Yet we monitor it much more frequently in a given year. Our main goal is to establish fruitful, long-term strategic alliances with suppliers, not just to procure for only a few years. In our supply chain contracts, we procure based on quarterly prices. In some contracts, we even procure at monthly revised prices. This system of frequent price updates came into our sector after the last global financial crisis. For instance, we procure using six-month-long contracts from Kordsa Global.[21] We have negotiations every six months based on certain price indices. After each negotiation, the price is fixed for the next six months.

Çağrı: Could we discuss your major raw materials?

Fatih: Our major raw material ingredient is *rubber*. There are two types: *natural* and *synthetic rubber*. Another major raw material is *carbon black*. We also have *textile* and *steel cords*. It is important to note that 60% of our raw materials are composed of rubber and carbon black. All of these raw materials are imported. Apart from natural rubber, the other materials are procured in a business-to-business trading setting. Hence, we manage them with *bilateral agreements and contracts* whose durations range from monthly to quarterly. For instance, carbon black is procured with quarterly priced contracts. The Bridgestone Singapore Natural Rubber Procurement Center handles all of the global natural rubber purchases for every Bridgestone facility. It acts like a broker for us. However, unlike a fast stock market broker, its lead time to process a purchase demand from us is rather long, at least fifteen days. Furthermore, the natural rubber we could use in our facility has to have a certain grade and quality. It also should come from an approved plantation. Hence, the *market liquidity* is an issue we have to manage carefully.

Çağrı: Could you elaborate on the natural rubber market development and your involvement in this market?

Fatih: Major producers are located in Malaysia, Thailand, Indonesia, and Vietnam. We generally trade with six-month advance forward contracts in this market. You have to purchase natural rubber via brokers in this market. We have been using the Singaporean market since 2004. The market was open to trading after 2004. Thus, not only the consumers of the natural rubber but also the traders who would like to make money have entered the market. Then we began seeing speculative price movements.

Çağrı: Surely these movements make your job much more complicated. This market has curious links with my previous work[22] that had been motivated by the wood pulp market. Since there are a number of regional markets for this wood pulp, such as Latin America or Scandinavia, there may be arbitrage opportunities. Yet the transportation lead times are long and cannot be easily managed. In such a market, how would you sell your wood pulp in a contractual setting where spot market trading also exists? In that work, we proposed an intelligent supply chain contract with a flexible option written into it that enhances the value of the contract. Price fluctuations in the market should be observed and in some cases spot market selling could be much more valuable.[23] In your case, there is also a similar commodity: *synthetic rubber*. How do these two materials compare? Are they really substitutes?

Fatih: Not really. A small percentage of our use can be switched to synthetic rubber, yet this cannot go all the way. You need to stop and use natural rubber. Therefore, you are exposed to the *market price risk*. On the other hand, there is research going on to reduce the dependence on natural rubber. This is a long technological development process. To mitigate the market price risk, we have developed a basic decision model that considers future price movements. In this model, we essentially created a few cases based on the price increase and decrease. In all these cases, we also created procurement scenarios that may say, for instance, purchase 20% now, 20% in the next month, and the rest in three months. Last year, we saved around US$5 million in the market using our decision model. Surely, we had good luck with the futures prices.

As a director, I would like my team to bridge the gap between academic studies and practice. I encourage them to apply new research ideas into our work. In another study, we developed a model to minimize the inventory level. Our model had to decide from which suppliers we should purchase (using a minimum of two suppliers) and how much we should allocate the inventory from these suppliers so that each supplier's monetary volume would be between 5 and 30% of the total portfolio and we hold twenty days' worth of inventory. In all these studies, the most difficult part was to predict the probabilities of the prices. Thus, we used certain heuristics. In our global supply chain management mindset, we try to establish a system that learns constantly and updates itself objectively. Otherwise, you are bound by the subjective opinions of the managers.

Çağrı: Sure. The know-how and know-why could be lost unless it is made explicit in any system. We conducted a project a few years ago. It was regarding the establishment of an effective know-how transfer mechanism of a major bus manufacturer to a Middle Eastern country. It is not at all an easy task. You need to design a system so that when key staff is absent, the system can continue operating in a resilient fashion. Otherwise, the system breaks down easily, as observed in many global examples.

Fatih: I totally agree. That is why we began to apply the SCOR model at Brisa. We wanted to prepare everything in a written format such that the system can continue by itself and also be sustainable. Let me continue on our natural rubber discussion. In monetary terms, it consists of 35% of the total raw material costs, yet it is 27% of the volume. In the Singapore market, we take as a reference price point the RSS3-type natural rubber. (See Figure 9.3 for the price evolution of natural rubber in this market between May 2007 and April 2012.) Once we submit our purchase

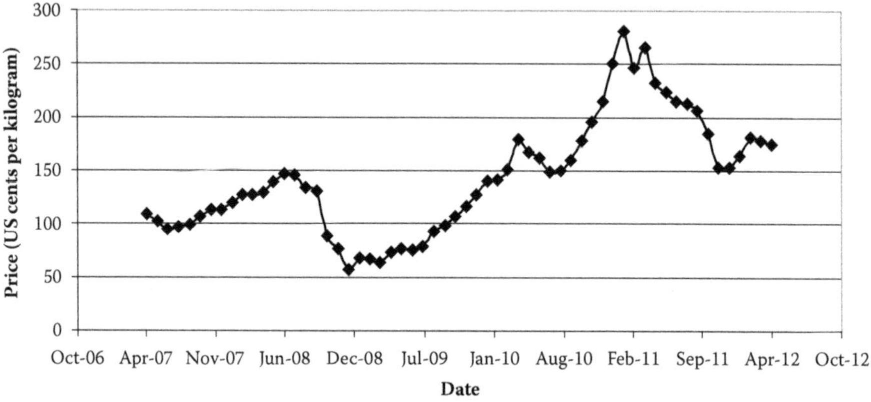

Figure 9.3 Singapore Commodity Exchange natural rubber price, May 2007 to April 2012. (From: http://www.indexmundi.com.)

order, the agent broker prepares the procurement with a forward contract. It takes up to twenty days for this process. When it is shipped, it takes around thirty days for ocean transportation. Since our purchases are conducted in a forward manner, we give orders and purchase for the previous three months due to long lead times. We do not have any control over the shipment time, which varies between thirty and forty days. It depends mostly on the producers' preparation time. We also consider the worst-case time for the agent broker, which is twenty days. We do this because natural rubber is a bulky commodity. When you hold it as inventory, it costs both in terms of money and space. Thus, we prefer to hold the inventory in the ship container, not in a warehouse, to avoid incurring extra holding costs. Hence, we aim to minimize our physical inventory along the supply chain network. Specifically, we have around twelve days' worth of physical inventory for natural rubber coming from the Far East. We really are on tenterhooks for natural rubber procurement. It is even troublesome for Bridgestone China, which is quite nearby the area and still operates with one month's worth of inventory. To secure a reliable supply of natural rubber, we use the worst-case scenarios for planning. In spite of this, a storm in the ocean and the Arab Spring events in Egypt—where we had the cargo transfer to a different ship—created worrisome days for us. Especially in Egypt, cargo ships transfer the containers to the feeder ships, which also add to the shipment times. Thus, we prefer direct shipments from the Far East. All of the cargo prices are calculated as Free on Board (FOB). We use three-month shipment contracts with our suppliers.

9.4 Risk Intelligent Supply Chain of Brisa

Çağrı: Let us start with the risk perception. How do you perceive supply chain risks at Brisa?

Fatih: We have had a risk management committee at Brisa since 2009. In that committee, our president, five vice presidents, and the risk manager work together. Since then, we have been working on assessing and developing strategies and action plans for all supply chain risks. Last year, we made an extensive study for this purpose where we prepared a *supply continuity plan*. We also determined our *key risk indicators (KRIs)*. In the meantime, we are beginning to examine the specific risks within the framework of the SCOR model and considering "source–make–deliver" cycles for each department. We have already finished the first phase of that study, and are now moving to the next phase.

İdil: We did a study with the risk management team to identify, assess, and manage supply chain risks at Brisa. We prepared a *supply chain risk map* for our company. (Refer to Figures 9.4 through 9.6 for details of the supply chain risk map.)

Çağrı: How is the supply chain risk management positioned within the overall corporate risk management framework?

İdil: We are positioned in a number of ways in corporate risk management. First, the *business continuity risk* is directly related to managing our global supply chain effectively. Second, the *social responsibility* and the *environmental risks* are on our radar screen. Third, the *reputational risk* affected by the brand value and the competitive actions are also related to the supply chain risk. Surely, the *price fluctuations of* the raw materials we consume have to be managed from both financial and supply chain risk perspectives.

Çağrı: How do you perceive the supply chain disruption risks at Brisa?

İdil: We have certain commitments in our supply chain. For instance, we do have to manage our contracts with our transportation subcontractors and raw material suppliers considering certain supply chain disruptions. Since we do have extensive export–import operations at Brisa, rules and regulations directly impact our supply chain operations.

Çağrı: How does the supply chain risk management process work?

İdil: We have an annual risk assessment workshop where all supply chain risks and action plans are decided. We have a *continuous monitoring system* that requires all Brisa departments to understand the types of risks that are gaining importance in the global supply chain network. These risks are reported to the committee in predetermined periods. In these workshops, first the risk sources are determined, and then the risks are clearly identified. Once this is done, the risks are given priorities, and thus the likelihood and the impact values can be determined. At this point, our

Risk Impact Level		Financial *Profit Loss* *(Million Tl)*	Reputation *Stakeholders:* *1. Shareholders, 2. Employees, 3. Customers, 4. Suppliers*	People *SHE* *(Inc. Natural Hazards)*	Environment
4	Very High	$x > 10$	• International impact • Total loss of public or stakeholder confidence • Serious damage on brand image	A big accident or a natural disaster that may cause the death of more than one person	Chemical material or waste release outside the company that is harmful for environment and health.
3	High	$10 > x > 6$	• Impact in one country • Long term confidence loss in a wide range of stakeholders/public. • Important damage on the brand image	A single accident that may cause the death of one individual	Chemical material or waste release inside the company that is harmful to environment and health
2	Medium	$6 > x > 2$	• Impact in one country • Confidence loss in limited stakeholders or region or in mid term. • Damage on brand image recoverable in one year	Accident that may cause working day loss of more than 20 days	Complaints about the noise and smoke kind of problems from the public or social networks
1	Low	$2 > x > 0.5$	• Local and limited impact • Confidence loss in small groups and/or in short term • Small impact on brand image	Accidents that may cause loss days	----------

Figure 9.4 Risk impact categories: financial; reputational; people; environmental; (safety, health, and environment (SHE)). (Courtesy of Brisa.)

LIKELIHOOD LEVEL	FREQUENCY	POSSIBILITY (%)
4	Every year	75–100
3	Once in 2 years	50–75
2	Once in 4 years	25–50
1	Once in more than 4 years	0–25

Figure 9.5 Risk likelihood levels. (Courtesy of Brisa.)

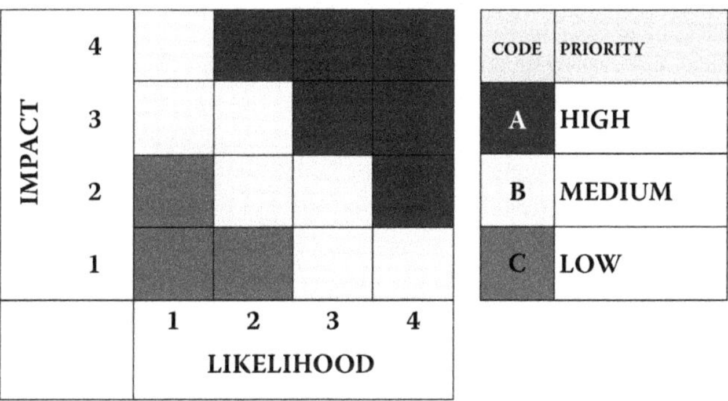

Figure 9.6 Risk categorization map. (Courtesy of Brisa.)

risk appetite helps to decide on the risk exposure. While the risk appetite is being decided, different methods are considered to manage these risks. If we do not want to take certain risks, we consider transferring them. Last, risk mitigation is done using various tools and techniques.

Çağrı: At that point, you need to decide on the method you would like to use to manage the risk intelligently. There are multiple alternatives. You might insure and hedge at the same time, or you might share the risk by using only supply contracts with flexible options.

İdil: Certainly. Our strategy forces us to think whether we could reduce the likelihood or the impact of a certain risk. Based on the risk type, our course of action varies. As in all cycles, we do have continuous monitoring and control over this risk management process.

Çağrı: In the risk management process, as we all know indicators play a critical role. How do you define and differentiate *key performance indicators (KPI)* and *key risk indicators (KRIs)* at Brisa?

Fatih: Thanks for this important question. Our risk department prepared a guideline on this topic as shown in Figure 9.7.

Çağrı: Sure. Suppose when your natural rubber shipment is traveling from the Far East and pirates attack[24] it around Somalia and you lose US$20 million.

KPI	KRI
It is used to measure an organization's development toward its preset goals. It is an "outcome." It does not indicate how much risk is taken to reach that outcome.	It is used to follow up the progress of risks over time. It indicates whether the operations are performed within the risk tolerances and might act as an early warning signal where there is a possibility of failure or an opportunity for improvement.

RISK TOLERANCE
It indicates the value of maximum acceptable level (limit) by management at which exceeding it must be considered as an alert situation or early warning signal for the defined risk. Criteria to determine risk tolerance:
(1) If exceeding the determined value directly causes an interruption of company operation or represents a situation that cannot be tolerable due to health, safety, environmental, reputational, or legal concerns, it must be considered a risk tolerance (e.g., zero A rank fire, zero fatal incident).
(2) If previous statistics for risk indicator show a nearly smooth trend and if deviations from the trend historically show an extraordinary situation at the same time, then the average or maximum value of the past trend can be considered a risk tolerance (e.g., some credit risk indicators, employee turnover indicators).
(3) If there is a clear target or boundary for the risk indicator set by partners, board members, or the executive committee, it must be considered risk tolerance (e.g., some budgeted financial loss or costs in each budgeting period).

Figure 9.7 Company guideline for KPI, KRI, and risk tolerance. (Courtesy of Brisa.)

That is the negative impact of a supply chain disruption risk that needs to be monitored through your KRI.

Fatih: If the piracy attacks increase around Somalia, that type of risk is considered in our KRI. Yet to be able to ship on time is considered in our KPI. We need to connect these two indicators since KRIs will give *early warning signals* to potential losses in certain KPIs. Another example is *fire risk.* If the number of people that smoke increases from 100 to 200, the fire risk increases, which needs to be reflected in our KRI. Or similarly, if the number of water valves, that are not working goes up, the fire risk increases. We do not claim that zero KPI means zero KRI. Two indicators need to be clearly defined, comprehended, and used.

Çağrı: Could you explain the most critical KPIs?

Fatih: From a supply chain perspective, there are a few KPIs we focus on day in, day out. One of these KPIs is using the *production capacity efficiently.* This is a must considering our continuous investments. The second one is enhancing the *sales forecasting accuracy.* In the last few years, we were not in the position we desired. This definitely affects the global supply chain operations from one end to the other.

9.5 Managing and Mitigating Major Supply Chain Risks

Çağrı: Now we can discuss in detail the major supply chain risks you are managing at Brisa.

İdil: The first supply chain risk we identified and analyzed is the *single-approved supplier risk*.[25] When we are bound by a single supplier for a given raw material, we are exposed to such risk. Even though the likelihood is not high, the potential financial impact can go up to US$10 million based on our estimates. We decide on the strategy as well as the course of action to manage this type of risk. That is, our strategy is to find and propose an alternative supplier. Our course of action is twofold: first, we need to determine the critical materials that are exposed to this risk, and then we establish a supply continuity plan including a backup supplier.

Fatih: We need to mention that we do not have that many critical raw materials. Yet we consider all of the materials from scratch from this perspective. Our team examines each material and finds out whether it is supplied by a single supplier. If it is, then the next step is to see if it is a global or local supplier. For global ones, they look for substitutes somewhere in the market. If there is an emergency, is it possible to use an unapproved supplier, or can we shorten the approval procedure or bypass a part of it? Also, we examine whether we can shift these material procurements from European suppliers to other locations. To mitigate such a risk, as a first operational hedge, *holding extra inventory* is the first method we considered. Furthermore, we prepared a *detailed geographic map* for the natural–synthetic rubber and carbon black. In that map, we located each of our suppliers in terms of prices, the percentage of our purchasing, the criticality of the items, and the potential problems. With such a map, we could calculate the potential financial loss and the impact on the working capital should such disruptions occur in the global supply chain network.

Çağrı: While doing this, have you also considered natural disaster risks such as earthquakes, tsunamis, and floods?

Fatih: We are working on them as we speak. We had some problems due to the Japanese Fukushima disaster as well as political incidents in Egypt. We know that a supplier in Egypt is favorable for us due to three days' lead time. Yet once we experience similar problems, it is fatal. Using a Far Eastern supplier is appropriate, but you have to hold extra inventory to mitigate the lead-time variability. And it is expensive. So there is an inherent trade-off that needs to be managed.

Çağrı: That is great. Because it seems that you have a constant learning mechanism where you update your beliefs and prior assumptions about potential risks as events unfold and surprise incidents occur along the global supply chain network.

Fatih: We did this the first time when we visited Bridgestone Global. We put forth this proposal so that we get priority over the global network in managing these types of idiosyncratic supply chain risks that we are exposed to.

Çağrı: I wonder how Bridgestone approaches these types of emerging global risks, especially after the Fukushima disaster.

Fatih: Our proposal included US$5 million savings in the working capital as well as around US$12–13 million savings in raw material procurement. Bridgestone is keener on the global impact of such emerging supply chain risks. Brisa might use 100,000 tons of such material, yet Bridgestone might consume much larger amounts such as 3 million tons in its global supply chain network. Therefore, the potential impact once quantified is definitely significant for them.

İdil: The second supply chain risk is the *failure in the supply continuity* or *reputational risk due to termination of supplier.* This risk is related to the potential bankruptcy of suppliers. This may cause up to four days of production interruption for us since there are potential alternative suppliers we can use. This risk is easier to manage than the single-approved supplier risk. For such a risk, we also establish a supply continuity plan where we determined the critical materials that are exposed to such risk. We also conduct commercial and financial analyses of such suppliers.

The third supply chain risk we manage is *delays in raw material delivery.* This is mainly related to transportation where the likelihood is higher with low impact. We also manage this risk through supply continuity plans and adjust our inventory holding policy for such materials.

Fatih: We aim to conduct statistical analyses of the materials for which shipment performance metrics fluctuate. We want to see the gap between the shipment promise and the reality, and the specific reasons for such gaps—whether due to the supplier, the road conditions, or Brisa. Such an analysis will also tell us for certain service levels what level of inventory we should hold to manage such variability. Thus, *holding extra inventory* is our major operational hedging tool to mitigate such a risk.

İdil: In the fourth supply chain risk, we focus on *curing bladder subcontractor risk.* Curing bladder is used in our press machines and being manufactured by a single supplier located in Düzce, Turkey. Should any problems occur in their facilities, our production could be disrupted, which could be very costly.

Çağrı: Does this company exclusively work for Brisa?

Fatih: Not exactly. This is a company in which we invested in its know-how generation for this production process. We are also working on a second alternative, yet it is not easy. If the production facility is burned down or destroyed by an earthquake, we definitely have a problem.

Çağrı: We know that Düzce lies on an active earthquake fault line in Turkey. We also witnessed a serious earthquake in November 1999, with a Richter scale of 7.2 in Düzce.[26]

Fatih: In fact, part of the factory collapsed during that earthquake. As of now, they have overhauled the entire building and made the necessary renovation. You may lose power in such an event and the production stops. In the worst case, we could move out to another location by transporting the molds. We are seeking such a location now. On the other hand, even if such an event can be insured, the potential sales loss is far greater than the financial loss. In two months, the sales loss in the market cannot be easily regained. So we cannot tolerate any production disruptions.

İdil: The fifth supply chain risk we manage is *single service supplier for domestic transportation*. This is due to failures in our single domestic transportation partner. Should such a disruption occur, we think that we might have some reputational loss but can resolve the issue within three days. This is a low-probability/low-impact type of risk. Our mitigation strategy is to continue deliveries with alternatives in case of emergencies.

Çağrı: Does this risk refer to just a blip in company operations or does it include the bankruptcy of that company as well?

Fatih: Any of those you mentioned. It can go bankrupt or cannot continue operations due to some other problems. In such an emergency, we could call upon a new transportation firm to continue deliveries. This may take a few days. Our transportation partner works with three groups of transportation vehicles. One is its own trucks. The second is a private exclusive cooperative. The third includes the trucks obtained in the market. Thus, if our partner fails, we do not lose most of the transportation capacity. Yet we lose all this coordination across the three groups. We know that taking over that process could be handled by Brisa with low financial impact.

İdil: In the worst case, we could use a cargo shipper to prevent any delivery disruptions in our supply chain. When we focus on the sixth supply chain risk, it is called the *excessive stock due to firm order cancellation*.

Once a firm cancels its order for the customized tires, we might end up with those tires and no potential buyers. For example, you may produce for a government tender. And the tender becomes void for some reason. You already have produced the tires, but now you cannot sell. This is the specific risk you are exposed to. We consider the impact and probability to be low; that is, one per year is quite a reasonable estimate for us. Our goal is to reduce the unsellable inventory to zero. So we need a different sales procedure to decide what actions we must take if we are working with a special customer. We collaborate with our sales team in this process.

Çağrı: Is there a specific case you lived through recently for this supply chain risk?

İdil: This happens often in the Middle Eastern markets. Our sales distributor wins a military tender in Algeria. Suppose they have to sell 5,000 special tires for that tender. They normally sell 100 units per month. In case the tender becomes void, the inventory of 5,000 tires can be sold in the next 50 months only selling 100 units a month. This surely adds to our holding cost. To mitigate such a risk, we closely collaborate with the sales team, constantly sharing *market intelligence* regarding such tenders. We might stop the production or reduce it partially considering the weak signals we obtain from the market. There is a second issue that complicates this process. Military tenders have strong penalties for tardiness. Thus, to prevent being late, you tend to produce tires earlier than usual. So timing of the production has to be carefully managed in this process.

Let us move on to the seventh supply chain risk we manage, which is called *delays in imported delivery.* Besides producing tires in Turkey, we also import tires from Bridgestone. These are tires with certain sizes and features we choose not to produce in our facility. The incoming shipment has a potential delay risk. This might cause both financial and reputational damage to us. Yet we can manage this risk mostly by having extra back-order tires at our distributors and dealers. Thus, the impact is low even though the frequency could be high. We aim to get shipments within the requested month. Hence, we request follow-up reports for the tires not shipped on time so that we can plan ahead with our dealers.

Çağrı: What is the transportation mode? Are you using ocean freight from Japan? How about European shipments?

İdil: We have been using ocean freight for imports from Japan. We also use ocean and land freights originating from Europe. For instance, shipments from Spain use the sea route, whereas Italian and Polish shipments use the land routes. It actually depends on where the tires will be shipped from. Sometimes in rare cases, we also use air freight, if necessary, from Europe.

Fatih: I need to make an important remark here. We focus on whether the supplier ships the items on time. Generally, the shipment delay does not occur in transit time; it is mainly due to the belated loading process at the origin.

İdil: For our eighth supply chain risk, we consider *delays in planned purchases excluding the raw material.* This risk covers the delays in our purchases in critical machinery and spare parts. These delays may affect the business continuity yet with shorter durations. Thus, it is regarded as low likelihood and impact risk. Nevertheless, we manage it using effective follow-up and warning systems with the related suppliers. We use automatic email notifications to the suppliers about the deadline of the delivery dates. Our current system sends messages one to two months ahead of the delivery dates. Now let us discuss our ninth supply chain risk, which is the *legal risk of the purchasing contracts.*

Fatih: After we overtook the OSAT, we saw that we had different types of contracts. These contracts had been designed with various perspectives of different departments that might enforce a variety of nondisclosure agreements. We decided to consolidate and standardize the agreements in terms of legal terms and conditions. After such a process, we also redesigned our IT system such that we could upload those contracts that contained the nondisclosure agreements. In the next phase of contracting, we now have a simpler and leaner process. Moreover, we are in the process of designing a completely new contracting module that will be overseen by our finance and legal teams. That will be the base of the *corporate contract management*[27] that also covers any type of contract, not only the supply chain contracts.

Çağrı: I would like to ask how the breach of contract risk[28] is being managed at Brisa. How is this risk being shared among partners?

Fatih: Our supply chain contracts—with suppliers apart from the raw material producers—are all designed such that we are secured in any case of breach of contract. We use advance payment guarantees in these contracts; thus, we are not exposed to the breach of contract risk. I need to note that we have not lived through any event of breach of contract with our suppliers at Brisa. When the job is done with a supplier, we do not pay 10% of the payment before the following six months. This time allows us to observe whether the job is properly done. We have some penalty terms in the case of delayed shipment. Yet we have not enforced such penalty costs until now.

Çağrı: It is great that you have not experienced any breach of contract event. Surely these events are rare in practice. It means that your relationships with suppliers are strategic alliances with a long-term mind-set.

Fatih: Sure. On the raw material procurement, we provide next few months' demand and negotiate on the prices. We do not have a long-term fixed-price contract. We generally use monthly orders but also present the long-term plans without commitment. For instance, with Kordsa Global, it is determined every six months. On the carbon black, we have three monthly contracts with monthly orders. For some special chemicals, we use monthly contracts. Our monthly allocations are generally flexible, not fixed beforehand.

Çağrı: Then fluctuations in the order frequency have to be managed by the suppliers. How do you manage that process with them?

Fatih: In the eyes of the suppliers, we have already built a great trust. The payments and ordering commitments all work perfectly with us. So the suppliers know that our variability is very low, and we can be trusted. Since we have a tedious approval process with suppliers, the tenures of suppliers are ten to twenty years. We call this relationship a *strategic partnership* rather than a customer–seller relationship. This allows us to be on the so-called VIP list in case of emergencies. So we are also a preferred buyer in the suppliers' world.

Çağrı: This is a critical topic. In the 1999 Taiwan earthquake, the semiconductor industry was badly affected. At that time, Apple and Dell were both supplied by similar firms in Taiwan. Apple had an intimate relationship with those firms and was able to shift some of the production capacity to itself, while Dell was somewhat late in this process. Thus, how firms behave in such crisis situations depends not only on themselves but also their global supply chain networks' capability and willingness to cooperate.[29]

Fatih: In our approach, for example, if the spot price is more favorable, we do not go and trade in the market. If we had a deal with a supplier, even if it is costly for us, we continue the procurement. On the other hand, if it is in the supplier's favor, we get a similar action from the supplier. So this works both ways.

For example, we had such an event in the Egypt case. The carbon black market was seriously affected in these events. There is a large producer in Egypt. The production had to stop, and the ports were unusable for a while. This caused a serious one-month delay. At that time, the European carbon black market was in a bottleneck due to closures of facilities in the previous crisis. Thus, firms were selling the material at premium prices at the time of scarcity. One European supplier, although it is an approved supplier—yet we did not have an established business with it—provided us the necessary carbon black in that crisis. This was due to our long-term *strategic partnership mind-set* we established with that supplier. We have currently around twenty to twenty-five suppliers with whom we are in such a strategic partnership.

İdil: The tenth supply chain risk we manage is *supplier risk*. This is related to the business continuity of a supplier. These are the suppliers that provide us spare parts and related services. The focus is mostly on the domestic suppliers. To manage this risk, we would have a follow-up commercial and financial analysis of the critical suppliers. Second, we would invoke an emergency plan in case of disruptions in any of these suppliers.

Our eleventh supply chain risk is called *raw material price fluctuations*. We know that unexpected fluctuations in the raw material prices negatively impact the expected profit.

Çağrı: This is essentially the market price risk in natural and synthetic rubber as well as the carbon black.

İdil: Sure. This is the only supply chain risk we categorize as high impact and high likelihood.

Fatih: The carbon black and synthetic rubber market prices are correlated to the oil price, which has high volatility nowadays. In our risk management strategy, we use forward contracts to hedge the price risk. Yet it is not sufficient. You cannot easily reflect the price increases in these materials in your products' prices. We have also developed an indicator that computes the *ratio of the total cost of raw materials to the total revenues.* We aim to continue with a certain target ratio. If we can reflect the price increases in our products, then this risk does not exist since the market is willing to accept it. In contrast, if you cannot reflect these changes, then the impact on the profit can be immense.

Çağrı: What are your mitigation strategies beyond the indicators?

İdil: First, we need to inform Brisa management for a timely action. Then we maximize the prevention against raw material price fluctuations. Additionally, we follow up on the market and price trends continuously. We exchange information with the suppliers and Bridgestone Japan.

Fatih: We were using a fixed annual price with our OE customers. We also designed an *RMI [raw material index] pricing scheme,* which is a pricing formula based on the price changes in the raw materials. We began using such a scheme with select large customers. This reflects the raw material price increases within the next one or two quarters. Surely, the OE customer segment is around 20% of our sales. We are working on using such an RMI pricing scheme with other customers, too.

Çağrı: This pricing scheme is intimately related to the sales contracts you design and manage with these customers.

Fatih: We have initiated a new model using *rolling forecasts.* It works as follows. We run the rolling forecasts every month based on our future expectations to observe the financial impact. This helps us decide our proactive actions for potential price changes and also the degree of reflection of these changes in the market. So, we could take actions much earlier and not just when the raw material prices have soared. This year when we decided on the budget in September, we were able to act early for price increases that would normally occur in May. This allowed us a much smoother price change in the market.

İdil: The twelfth supply chain risk we manage is *information security risk.* It is defined as the leakage of confidential documents to the outside. We use levers such as managerial controls over following up, training, and obeying company confidentiality policies. Besides, every year Brisa conducts *information security* training. We also use an online training portal called *Briport.* Surely, this risk is closely related to the trust between the company and its employees. This risk is more pronounced in the technology development teams that use our confidential mixture prescriptions. (Please see Table 9.1 to examine all of the major supply chain risks we have discussed up to now and the relevant details in a summary.)

Table 9.1 Major Supply Chain Risks and Mitigation Strategies

Risk	*Definition*	*Impact*	*Likelihood*	*Target/Strategy*
Single (Approved) Supplier Risk	Failure in supply continuity of raw materials for which there is a single approved source for Brisa	High financial loss due to long-term production interruption	Low probability	Find and propose alternative supplier
Supplier Risk (Raw Material)	Failure in supply continuity or reputational risk due to termination of supplier	0–4 days production interruption	Low probability	Supply continuity
Delays in Raw Material Delivery	Delays in raw material delivery may cause production interruption	0–3 days production loss	Every year	Supply continuity
Curing Bladder Subcontractor Risk	Termination of current sub-contractor for any reason will cause a significant production interruption	Nearly 2 months' production interruption	Low probability	Supply continuity
Single-Service Supplier for Domestic Transportation	Failure in domestic transportation due to problems in supplier	Reputational loss due to approximately 0–3 days delivery delays	Low probability	Delivery continuation
Excessive Stock Due to Firm Order Cancellation	Excessive stock due to firm order cancellation for customer-specific products	Depends on the tire quantity cancelled	Every year	0 excessive (unsellable) stock
Delays in Imported Tires Delivery	Financial or reputational loss due to supply risk + excessive stock risk	Low probability of order cancellation	Every month	Shipment within requested month

(continued)

Table 9.1 Major Supply Chain Risks and Mitigation Strategies (continued)

Risk	*Definition*	*Impact*	*Likelihood*	*Target/Strategy*
Delays in Planned Purchases (Excluding Raw Material)	Business interruption or scrap due to delays of critical machine purchases	Short-term interruptions, very low impact	Low probability	Minimizing delays by effective follow-up and warning systems
Contract Risk	Legal risks of the purchasing contracts	Almost no penalty cost to Brisa are expected	Low probability	Minimizing the contract gaps and risks Standardization in contract management
Supplier Risk (Excluding Raw Material)	Failure in supply continuity or reputational risk due to termination of supplier	Low reputational and financial impact	Low probability	Supply continuity
Raw Material Price Fluctuations	Unexpected fluctuations in raw material price, negatively impacts expected profit	High financial loss due to unexpected raw material price fluctuations	Fluctuations every year	Inform Brisa management team for timely action Maximize the prevention against raw material price fluctuations
Information Security Risk	Leakage of confidential documents (supplier info, etc.) to outside	Low reputational impact	Moderate probability	Minimum information security breakdown

Source: Brisa.

Çağrı: It is great to see that you already identified the most critical supply chain risks and decided on how to manage them. I want to learn how you approach the risk assessment.

Fatih: We both use objective and subjective methods. Historical data are used for statistical analysis. But we also use intuitive heuristics that have been tested for many years at Brisa.

Çağrı: How about natural disasters in your global supply chain network? How do you manage them?

Fatih: Since we believe that some of our raw materials in the Far East may have disruptions, we plan to have a dual-sourcing strategy composed of a near source and a faraway source. For natural disasters such as earthquake, flooding, or even fire, our company deals with them within business continuity planning.

Çağrı: On the other hand, there are also *environmental risks* nowadays gaining more attention. Certificates such as ISO-14000 are being awarded. This is intimately related to climate, energy, and sustainability. Carbon dioxide emissions need to be measured, and the risks created by changes in climate and energy use have to be managed. These are also related to the regulatory risks created by the Turkish government and the European Union. What is Brisa's approach in this realm of risks?

Fatih: If the items we procure require a certain certificate due to regulations, we ask for these to conduct business with these suppliers. If it is not a requirement but the supplier has such a document, it is on our wish, not must, list. On the other hand, we also have been dealing with different risks in the context of *corporate social responsibility.* Together with Bridgestone, we prepared a survey of 400 questions including *child labor* and used it with our suppliers. A few years ago, we did a self-assessment and prepared an improvement plan. Next year, we will share this information with our suppliers and request similar action plans. We also support *fair trade* and *transparency,* which includes our system of sharing with the media and society.

On the carbon emission side, we developed our own system to measure CO_2 consumption in our supply chain activities. We also agreed to provide data to the CDP [Carbon Disclosure Project].[30] We became an executive member of the CDP. We are not at the stage of measuring the emissions for our suppliers' logistics activities. We just measure the emissions in the production processes. Discussing with the Turkish maritime and the general directorate of highways, we developed a method that uses the vehicle type, mileage, and type of fuel to calculate the emissions. We did this for all the logistics activities, that is, incoming and outgoing shipments. Our current goal is to decrease emissions by using sea freight and railroad more often. Brisa is also a member of the Global Reporting Initiative,[31] a nonprofit organization.

Çağrı: How widespread is railroad use in your supply chain transportation?

Fatih: We use it for both domestic and foreign transportation. For example, we use railroad transportation to Syria or Hungary and sometimes to the Czech Republic. When its costs are similar to highway transportation, it is not preferred due to its relative inconvenience. Also, the lead times are generally longer and not that flexible compared with highway transportation.

9.6 Regulatory and Political Risks and Brisa's Risk Intelligent Response

Çağrı: When you consider the regulatory risks, are there specific risks that have affected you recently?

Fatih: We have been impacted by *EU environmental regulations*. For example, two years ago, we had to completely change the type of oil used in production. We switched to a different aromatic oil. This was a costly process. Also, under the EU regulation called REACH [Registration, Evaluation, Authorization, and Restriction of Chemicals], we need to be careful not to use carcinogenic materials or those that are harmful to the environment and health. This has to be proven for each material we use. Our global suppliers are carefully following these regulations. On the other hand, the Turkish government enacted a rule that is stricter than the EU regulation. They control the materials imported to Turkey much more severely. However, I need to say that since these regulations are enforced in a longer time frame such as two to three years, adaptation is much easier. We have not experienced any issues in the supply and business continuity apart from the cost effects.

Çağrı: How about the political risks along the global supply chain network? You mentioned about the events in Egypt. Could you elaborate on these risks and your approach to managing them?

Fatih: Our risk department publishes monthly reports that cover country risks. In these reports, we have all types of risks covered for specific countries that Brisa works with, such as economic, political, seasonal, and natural disasters. Once our supplier in Egypt stated that production was disrupted, we first asked the supplier to secure part of the inventory for us. Then, we went to a different supplier in India. Surely, the lead times were very different: three days from Egypt; thirty-five days from India. On top of that, since the material we would get during this time seemed to be insufficient, we changed our specs for some of the products. We switched to other substitute products by adjusting the production plan. Besides all these, we got commitment from another European supplier to provide three to five days" worth of material, which is one of our strategic partners, in case of emergency.

Çağrı: Very interesting. How did you find the Indian supplier in such a crisis?

Fatih: It was already an approved supplier in our global supply chain network. We were not procuring from that supplier due to higher cost and longer lead time. When it was necessary, we did it. In the overall process, we worked in cooperation with Bridgestone. Since the global carbon black procurement is managed by Bridgestone, events in Egypt affected not only our facility but also other global facilities. In the end, events did not last long, and we did not have to use the European supplier. Using the Indian supplier and the actions we took in our production process mitigated our problems. Nowadays, we procure not only from the Egyptian supplier but also the Indian supplier. Thus, we are using a dual-sourcing strategy for carbon black. The Indian supplier is not a panacea due to its long lead time, yet having it in our supplier portfolio is important to hedge similar risks.

Çağrı: Great story. How about the Japanese earthquake–tsunami and the Fukushima disaster? How did it affect Brisa?

Fatih: I am glad to say that the impact was minimal. Surely Japanese facilities were affected. There was one particular material we were procuring from Japan. With the coordination of Bridgestone, we switched to a U.S. supplier. We also did some *production prescription changes* and used a *substitute material* with the same specs in our production. All of our supplier selection is done in cooperation with Bridgestone. For instance, in August 2011, we experienced another event in Malaysia. The material we would procure had not been loaded. Our production would have been disrupted in such a case. To be able to find the same material, our team worked together with Bridgestone and were able to locate another European supplier. After receiving approval from Bridgestone, we solved this issue by procuring from the new supplier.

Çağrı: So it is clear that working in close collaboration with Bridgestone to manage global suppliers has a positive impact.

Fatih: Certainly. Working in a *global network of suppliers* is advantageous. For example, we frequently have problems in natural rubber procurement due to *floods*[32] caused by monsoon rains. In such cases, loadings and shipments may be delayed. Then, we can use the inventories at the European facilities. In a similar vein, when a European facility experiences such problems, we direct our materials to Europe. Thus, our global supply chain network creates flexibility and adaptability.

Çağrı: In essence, you are using a global transshipment strategy between facilities to mitigate any potential reputational risks. I think this is critical not only for Brisa but also for Bridgestone. Can we now discuss the financial hedging strategy of Brisa in managing supply chain risks? Besides forward contracts trading, what other tools does Brisa use?

Fatih: We have business continuity insurance due to supplier misconduct and for the materials in transit. But we have not obtained any insurance for the supply risk yet.

Çağrı: Do you use *risk pooling* at Brisa in any of the supply chain processes?

Fatih*:* Sure. When the products go to the dealer inventory, our IT system can observe the sum of the total inventory in a consolidated fashion. We just allow dealers to place their inventory in the common pool of which they do not want to sell themselves. This inventory can be seen by any dealer in our network and can change hands based on dealer demand. The second step is to manage this process ourselves and decide on the optimal transshipment levels across the dealers. Still, overstock and understock events[33] today cannot be fully mitigated in our system. Nevertheless, sharing the inventory information in the pool creates an extra flexibility in dealers' optimal decisions. We should not forget that the dealers are independent corporate entities.

Çağrı: For such a system to work, when dealers are independent entities, you need an incentive mechanism so that profits are shared fairly in the system. This may encourage them for the good of the system as a whole.

Fatih: You are right. But sometimes when a dealer cannot sell tires, maybe it is to its benefit not to sell.

Çağrı: Can the dealers also sell other brands under the same roof?

Fatih: They do sell whatever brand they like. After the *block exemption regulation,* there is no obligation to sell a single brand. We cannot contract with them in such a way that they become exclusive Brisa dealers. This is against the competition act.

Çağrı: Could you please elaborate on the block exemption regulation?

Fatih: It is another EU regulation that has been in place for two years. It states that you cannot put a precondition on any dealer to just sell your own products. This is in fact valid for automobiles. It operates now by mutual agreements such that a Ford dealer sells only Ford vehicles. Normally, there is no legal sanction.

Çağrı: What happened after this regulation? Did the dealers ask to sell your competitors' products?

Fatih: Most of the time, we entered into competitors' shelves. It worked favorably for us. Our dealers became free to sell any tire they desired. Yet we work toward making our products more attractive to sell. We provide *financial incentives* for selling our tires.

9.7 Learning from Black and Gray Swans in Swan Lake

Çağrı: Considering *black swan* risks, that is, having low-probability/high-impact risks, what is your approach to managing these unknown events?

Fatih: To that end, we have done SWOT analyses companywide and for the supply chain operations. Once we did this, we identified the strengths, weaknesses, threats, and opportunities and acted accordingly.[34]

Çağrı: On the other hand, an indispensible part of a solid risk management system is learning from failures as well as near-miss events, that is, gray swans. These near-miss events are generally studied in corporate accidents in different industries such as chemicals, automotive, oil, and health care. In your supply chain framework, are you learning from these types of events? If yes, can you explain briefly how?

Fatih: After near-miss events, we sit down and conduct a *postmortem analysis*.[35] We sometimes make urgent decisions that are mostly improvised to prevent future major disruptions. Sometimes, we are late in updating our views and making decisions. But now after having prepared a detailed business continuity plan, we know what to do, how to take action, and how to take the necessary precautions for the future.

9.8 Recruiting, Developing, and Retaining Supply Chain Talent

Çağrı: I want to learn how Brisa manages the *talent supply chain*. You are working with individuals with a diversity of perceptions. Do you have a winning formula or golden rules to recruit and retain great talent?

Fatih: In Turkey, the supply chain is still perceived as only transportation in some firms of our industry. First, the comprehensive mind-set for supply chain management has to be developed in the industry. This will create extra synergy among the different players in the industry. Thus, the most important characteristics of new talent for supply chain management is having a *capacity to visualize the big picture* and manage it accordingly.

At Brisa, we took this important step ourselves and gathered five different functions under the roof of supply chain management. This allowed different types of teams such as procurement to come together and discuss issues with sales teams. This platform enabled people to see the big picture. Besides, I think giving people different titles on top of what they do in the firm is also effective. For example, Ms. İdil Ertürk is our supply chain risk coordinator apart from her responsibility of production planning. Another team member is the IT or the customer service representative coordinator. We call this, in managerial jargon, job enrichment. This is a formula we use to bring in talented people who would love to enrich themselves with extra responsibilities.

Çağrı: So you believe that the big-picture mind-set is critical without losing the operational details.

Fatih: One should be aware of the big picture without losing sight of the details. Any failure, defect, or slowing down in the operations may be costly. One should be aware of those costs and their consequences. Therefore, we need people who are sensitive to what they do at the company.

Çağrı: Great. You are aware of the *poka-yoke* concept. These are systems and tools designed to avoid inadvertent errors. Surely you should have these tools and methods in your production process. Have you thought about using them in the supply chain risk management process?

Fatih: Of course. These could be designed at operational levels. The execution of operations without any errors is a must, yet a more important necessity is to manage operations much more effectively at a macro level. As mentioned, we began our journey with the SCOR model and work on redesigning all our processes. These types of tools will be added into the system. As time moves on, we may need to replan the necessities that evolve.

Çağrı: How do you plan to improve *supply chain risk awareness* at Brisa as you go along?

Fatih: This is very important. As Brisa becomes more aware of the internal and external supply chain risks, how to best manage them becomes the issue. We need to be better at learning new tools and concepts as well as unlearning and forgetting the old stuff. For instance, the KRI concept is new for us. There is a significant managerial pressure to decrease inventory levels at each stage of the supply chain in order to reduce the working capital. However, leaner supply chains are more prone to disruption risks. How to balance these two conflicting goals is our challenge at this point. For this purpose, we are working on developing a novel method to manage the inventory levels in the presence of potential supply continuity risk. As you will agree, there are no off-the-shelf solutions for such nontrivial problems. You need a strong team of people who have relevant backgrounds for developing and improving such systems.

9.9 Managing Business and Supply Continuity

Çağrı: How does the *supply continuity plan* work at Brisa?

İdil: Let me start with the main purpose of the supply continuity plan. We aim to minimize business losses and reputational harm by focusing on major issues during supply risk to set a clear and systematic process for searching out alternative supply solutions. While doing this, we focus on well-categorized supply problems to be prepared proactively from problem initiation until the risk committee decision-making stage. The scope of supply chain continuity covers all supply-related risks, which are defined as a *delay, stop,* or *insufficiency in product supply to customers* that might cause the breakdown of our commitments and thus financial or reputational losses for our company. First, we determined that supply risks are occurring in products, raw materials, and/or facilities. For each one of them, we observed that three root causes exist: *quantity, quality,* and *transportation.* We developed a number of different scenarios that are potentially observable for each cause. For example, in the product category, the quantity may be a root cause if Brisa has a temporary or permanent shutdown. In such a case, we cannot supply the required amount of products to the market. We could also have insufficient production attainment or insufficient product quantity as two other scenarios. In a similar vein, for quality and transportation root causes, we would observe a product quality incident or a product logistics problem.

In these types of events, we also identified *risk initiators,* those people who first identify the risks and inform others at Brisa.[36] These initiators are the responsible people who coordinate with others to rectify the issues. We also determined the communication medium that would be used in such a case. Last, based on the severity and likelihood of the risks, we categorized them into two major classes: *critical* and *noncritical.* Let me elaborate on these classes and how they are being managed.

A critical risk is defined as the type of risk that causes a significant delay, stop, or insufficiency in product supply that might cause a breakdown of contract commitment for OE customers or might cause a high financial and reputational loss to Brisa. A noncritical risk is temporary and can be tolerated with limited loss. Once a critical risk occurs, the risk committee must be called for an urgent meeting. And all necessary preparations have to be done by the related departments, and these preparations need to be consolidated by the product supply planning manager before the meeting. This is an emergency situation. In contrast, if it is a noncritical risk, then to take the required actions it is sufficient to communicate with the customer. Now how do we decide if a certain supply risk is critical? We have a number of questions in Table 9.2. If we get yes as an answer to any one of them, we consider that risk to be critical.

Table 9.2 Questions Used to Identify the Criticality of Supply Risk

Does this risk affect any of the critical OE customers' demand?
Does it cause a production stop at the OE customers' plants?
Does it have a direct effect on the brand image?
Does it cause a production stop of all of our products?
Does it cause a delivery disruption of more than two days?
Does it cause a production stop in a specific product category?

Source: Brisa.

If any of these questions lead us to a critical risk, our team member who is responsible for the supply chain management risk (currently it is the product supply planning manager) calls for an urgent meeting with each of the risk committee members. The supply risk committee is composed of our chief executive officer (CEO), chief financial officer, chief security officer, chief marketing officer, chief technology officer, supply chain management director, risk manager, and product supply planning manager.

Çağrı: The committee seems to have a diverse set of people. How do you manage the information flow internally and externally?

İdil: The product supply planning manager coordinates this information flow. Based on the criticality of the risk previously discussed, internal and external stakeholders are provided information. For example, our chief sales officer contacts the customers when necessary. The CEO may inform the media and the board. On the other hand, our corporate communication manager communicates with the public relations agency in case we have reputational issues.

Çağrı: Could you elaborate on the details of the root causes and how you incorporate them into your decision making?

İdil: Certainly. We stated that the preparedness and response components for important issues can clarify our actions. Therefore, we have designed information flowcharts based on the root causes in the supply continuity plan. Once a root cause occurs, who will do what? Who should be informed? What are the courses of actions? We know that we all have very experienced staff in our supply chain team at Brisa. Yet when someone inexperienced arrives, we need to be able to train that person in a standard fashion. We prepared these information flowcharts having such cases in mind.

We also need sustainability in our own system and its management. At a very early stage, we aim to address a *supply continuity risk* event. Therefore, the system works like an *early warning mechanism* for us. For instance, if a case turns out to be a critical risk as previously defined, we need to inform the risk committee within one day. There are nine root causes to be identified. However, we decided that any quality or transportation issue can cause a quantity problem; that is, we cannot deliver the required quantity to the customer. The supply is disrupted. Thus, we concurred that the most critical root cause is the product quantity. Every other root cause such as the quality and transportation triggers the product quantity root cause. Now if you allow me, I will elaborate on the information flow for the *product quantity*.

When one first observes a product quantity problem, that person informs the sales teams immediately as an early warning. At this point, risk bells do not toll yet. We ask whether the sales loss can be tolerated. If the dealers can handle the situation with the balance due, you accept the determined sales loss and stop. If you cannot tolerate the sales loss, can you find the necessary amount from another channel, or can you shift the products from some other customer? If this does not work, then one considers the following options in a sequential manner. If one cannot be used, you move on to the next option:

- Can you revise the production or the import plans?
- Can you provide a substitute product, for instance a Lassa tire instead of a Bridgestone tire?
- Can you import tires from an alternative source?
- Can you bring tires from customers' warehouses?

If none of these options are feasible and cost-effective, the responsible supply chain management risk person has to be informed. Then the risk criticality needs to be determined by asking the questions in Table 9.2. If it is a noncritical risk, informing the sales group is sufficient. If it is critical, then informing the risk committee is needed. Then the risk committee takes the necessary follow-up decisions. Surely, once the risk is identified, the process does not end. Everyone involved needs to take actions to mitigate the risk. If the issue is due to a machine breakdown in the production line, people begin resolving the issue as soon as possible. While the problem is being resolved, we—the supply chain team together with the sales team—are in close contact with the customer to manage this process.

Now, let us discuss the raw material quantity root cause and how it is being handled. This also triggers our product quantity root cause. This process is managed by the procurement team. When a raw material quantity problem occurs, we start with the first question to see whether the current supplier can provide the required quantity within the acceptable time period. If the answer is yes, one takes the necessary actions and the problem is resolved. If the answer is no, then we have the following options to consider sequentially:

- Is the raw material allocation performed by Bridgestone Global?
  - Can this allocation plan be revised?
- Is there an approved alternative supplier?
  - Can this supplier provide the required material within the time needed?
- Can the industrial engineering department change the daily production schedule and gain time by production postponement of that particular tire?
- If none of these options can be used, then inform the product supply planning manager for necessary revisions in the sales plan and initiate the *product quantity* process.

Çağrı: How about the *quality-related root causes*? How are they being managed?

İdil: When a product quality problem occurs, again we first examine whether the sales loss can be tolerated. If it is a minor sales loss, we accept the determined sales loss and continue business as usual. If it cannot be tolerated, then the decision flowchart is initiated. The suspected tires are quarantined in our system. We ask the following questions:

- Is the technology investigation adequate to release the quarantined tires to sales? In other words, is the quality problem suspicion false? If yes, take the necessary actions and the problem is resolved.
- If the first option is not possible, can the product allocation be shifted from other channels? If we sell the same tire to our dealers and the OE customers as the sole supplier, we should treat the issue first with the OE customers. To avoid stopping their production lines, we may have to put dealers on hold.

Çağrı: Can the priority be reversed? That is, you give priority to the dealers instead of OE customers.

İdil: In general, our critical customers are the OEMs. We cannot risk stopping their production lines. You can work with dealers using backorders, which is not possible with OE customers. Tires have to be assembled onto cars, and the cars need to leave the assembly lines. We are dealing with every OEM in Turkey. Remember, if a risk affects the OE customer, we label it as critical. Once it is noncritical, you have more degrees of freedom for risk mitigation. If we have a supply shortage from such a common set of products or the product, we generally assign the utmost priority to the customer with the highest market share for that product. In other words, if we are the only supplier of that company for that product, we try not to harm their production since they do not have another supplier.

Çağrı: How about the *transportation root cause*? How is it being managed in your supply continuity plan?

İdil: This root cause is mostly applicable in product and raw material transportation processes. Once a problem occurs, we first examine whether the delivery date proposed by the supplier is acceptable. If it is acceptable, we are done. If not, we seek to use an alternative delivery method such as air freight. We similarly ask:

- Can we revise the production schedule?
- Can we supply from another source?

If none of these questions can be answered positively, then the issue becomes one of higher priority, that is, a product quantity issue. Assume the raw material does not arrive in the next fifteen days. If you have a sufficient amount of inventory on hand to mitigate this late shipment, then it is a tolerable risk and does not negatively affect the supply continuity.

Çağrı: Great. Thanks for explaining the details of the supply continuity planning process. Since it is also written down, it becomes part of the Brisa know-how. You can now work on improving them.

İdil: It was important that we put down all the details in a formal process. Who would work with whom at what level? This supply continuity plan has been published as a company policy. It is now a part of our company rules. We are constantly revising it as days bring new issues and problems to be addressed.

Moreover, we also discussed with our risk management group to develop a new portal that will manage the *supply chain risk event reporting* in real-time. We already have such a system working actively in our SAP infrastructure. When one reports an issue for any module of SAP, that issue is logged into a portal and the related staff is being informed about it. Then the necessary rectification process takes place and courses of actions are reported at the portal. Our IT team is pretty experienced in managing such a portal. So we aim to develop a completely new *risk portal* for Brisa. Such a portal will also act as a medium to preserve the *corporate risk memory.* We already finished the conceptual design and determined the details of the areas to be covered and the individuals involved.

Çağrı: Essentially, this portal can act as a *risk logbook* for Brisa.

İdil: Indeed. Now we are at the IT software development stage. In the end, we will have a portal so that authorized staff will be able to report risk events in real-time, which will allow us to be more proactive in managing and mitigating supply chain risks.

Çağrı: Also, this system can act as an early warning mechanism for such risks that are hard to decipher lingering in the supply chain network. When you collect data in such a fashion for a few years, it will be of great use for further analysis and managerial insight.

İdil: Sure. Everyone related will be informed in real-time automatically. We will be able to monitor the costs of mitigation in the process that are also very important in our decision making. And we will be seeing the frequency of the risk events as well as their financial impact on our supply chain operations.

Çağrı: The faster you collect such data, the better and more effective the system becomes.

9.10 Envisioning the Future

Çağrı: How do you envision supply chain risk management at Brisa in the next few years? What are the issues you are most concerned with?

Fatih: In my view, risk is growing in importance as we go into the uncertain future. We are now focusing on a variety of risks that never existed before. It is clear that not firms but supply chains are competing with each other today. Quality has reached a certain level and has lost its significance for differentiation. In such a world, your speed, agility, flexibility, and management of the risks become essential. You need to be managing them extraordinarily well. You need to plan for any contingency. So risk management teams will gain more importance in the future. Besides, the Turkish government has imposed a rule within the trade law to present a bimonthly risk report to the board of company management.

Çağrı: Is there anything you can think of to modify or improve in your risk management process?

Fatih: We have already formed a risk committee as mentioned. Even if you try to cover every type of risk along the global supply chain network, there are some blind spots as well as black swans you cannot uncover, such as the case seen in Malaysia. We need better and more timely visibility along the supply chain network. I think we can develop such a system that would bring almost everything on one screen in real-time.

Çağrı: Like a *risk dashboard?*

Fatih: Sure. This kind of a dashboard requires a solid IT infrastructure. Although our infrastructure is working fine, it needs to be fully *integrated with our suppliers*. This is not a trivial task. Sometimes we cannot learn on time whether product loading is done even if we run after the supplier and the shipping company. Some of the suppliers are not ready for such tasks yet. However, we are developing new models to attain better and more timely visibility along the supply chain. For instance, we have just begun using a new application for customs processing. Previously, work orders were transmitted via phone calls. Now we have developed a module on the Web that is integrated with SAP. Once an order goes to the customs clearance company stating that the products have arrived, an email is transmitted automatically to the customs clearance company and the shipping firm. So now we can monitor the location of the container and its status. We can observe in real-time whether the container is in transit or at a particular warehouse. This allows us to manage the shipments more effectively.

9.11 The I-Quartet Model of Brisa

These lively conversations outline Brisa's approach to risk intelligent supply chains. Table 9.3 summarizes Brisa's I-Quartet model for the risk intelligent supply chain. From this summary, we can infer that Brisa's risk intelligent supply chain seems to be much more visible through the Inquirer and Ingenious roles. However, the

Table 9.3 Brisa's I-Quartet Model

Integrator	*Inquirer*	*Improviser*	*Ingenious*

• Global integrated supply chain network for critical raw materials such as natural rubber and carbon black
• Strategic partnership with 20–25 global suppliers
• Orchestrating global and local stakeholders seamlessly
	• Supply chain contract price revisions based on market intelligence (monthly to quarterly contracts)
	• Close follow-up of global commodity markets, constant learning and updating beliefs
	• Use of KRIs as early warning signals that are tightly coupled with KPIs
	• Use of market intelligence sharing among sales teams to manage order cancellation risk
	• Use of financial ratios as an early risk indicator for market price risk (e.g., ratio of the total cost of raw materials to the total revenues)
	• Conduct postmortem analysis after near misses and failures for smart learning
	• Use of detailed supply continuity planning to enable early warning signals
		• Political risk management (e.g., managing political incidents and disruptions in Egypt with vigilance and improvisation)
		• Improvisation plays a great role at Brisa for decision making in times of crises (e.g., Malaysian case)
		• Designing and using supply chain contracts with flexible and adaptable terms and conditions
			• Procurement decision model to manage the market price risk for natural rubber using forward contracts (US$5 million savings realized)
			• Intelligence in managing the 12 key supply chain risks
			• Use of various operational hedging tools such as
			  • Production prescription changes
			  • Substitute products
			  • Inventory and capacity pooling
			• Financial hedging with business continuity insurance for supplier misconduct
			• Using risk initiators to identify and communicate risks similar to extremophile response
			• Dual-sourcing strategy (near and faraway sources)

other roles of supply chain risk intelligence (i.e., Integrator and Improviser) do play invaluable roles in managing the global supply chain network. The amalgamation of the four roles in the I-Quartet model creates Brisa's thriving leadership in the age of fragility.

Now, as our final case study, let us move to a services company, Turkish Airlines, and its unique brand AnadoluJet. Using Boeing aircrafts in its fleet, AnadoluJet can be considered as an indirect customer of both Kordsa Global and Brisa. Hence, it is part of the global tire supply chain network—as a final user of the tires.

Chapter 10

AnadoluJet: Maestro over the Clouds

As you become the first address to be known, you start gaining traffic; once traffic is established, everything boils down to how well your model is built. Thus most failure is due to a lack of concrete model but not lack of proper marketing.

—Sami Alan, Senior Vice President of Regional Flights for AnadoluJet (Turkish Airlines)

10.1 Background of the Turkish Airline Industry

Up until November 2003, the airline industry in Turkey was regulated by the approximately 97% (now approximately 49%) state-owned enterprise Turkish Airlines (THY[1]) operating as the sole carrier. Apart from Turkish Airlines, for the incoming foreign tourism market, local and international charter carriers were in charge, targeting especially Antalya.[2] In November 2003, domestic flights were not as scattered as they are now, covering only primary routes such as İstanbul–Ankara, İstanbul–Antalya, and İstanbul–İzmir. These routes had a fixed economy class price frequently adjusted due to hyperinflation until 2003. As of the end of 2003, a one-way domestic price was a steep 149 TL and in August 2012 was around 200 TL. Demand was consequently low at these high prices; thus, the flights' load factor was not enough to increase frequency, which is also one of the main drivers—other than price—of attracting new customers.

In November 2003, the Turkish finance and transportation ministries took a radical step toward deregulation. Special incentives were provided to lower the domestic ticket prices. Specifically, two taxes—the *special operational tax* and the *earthquake tax*—levied after the August 1999 İstanbul earthquake were removed. These taxes amounted to 40 TL, that is, a 25% addition to the individual ticket prices.

Once the industry was deregulated, Onur Air was first to begin operations in November 2003. In April 2004, Atlas Jet became the second player, and Pegasus Airlines introduced itself to the market in 2005. Onur Air and Atlas Jet, operating out of İstanbul Atatürk Airport as their hub, began flying to several major cities including but not limited to Antalya, Bodrum, Adana, and Trabzon. On the other hand, Pegasus Airlines decided that its hub would be the secondary airport of İstanbul located on the Asian side: Sabiha Gökçen. Despite having a relatively smaller capacity than the İstanbul Atatürk Airport, Sabiha Gökçen represented a great potential as Atatürk was rapidly reaching capacity. And it was obvious that Sabiha Gökçen was going to enjoy spill traffic, coupled with the fact that the residents of the Asian side of İstanbul would be hesitant to go to Atatürk, because of the heavy bridge traffic, and just fly from Sabiha Gökçen. Later in November 2006, two more airlines made İzmir their main hub. The first one was SunExpress, a brand jointly founded in October 1989 by Turkish Airlines and Lufthansa as a charter firm targeting tourism, especially from Europe to Antalya. As İzmir became mature as the third largest city suited for international and domestic flights, SunExpress leveraged the potential of İzmir with so-called seat-only business. It did not take too long for İzair to enter; it was a start-up of the İzmir Chamber of Commerce, which then sold major stakes to Pegasus two years later. In early 2008, there was also a short-lived operation called FlyAir out of Trabzon, which traveled just between Ankara and İstanbul. Although the initial performance was far from miserable, it quickly bankrupted.

We need to note that deregulation and the rush of new firms increased the domestic passenger numbers from 8.7 million to 13–14 million just in one year, 2003–2004. Six years later, in 2010, 50 million domestic passengers flew over the skies of Turkey.

This phenomenal domestic transportation growth cannot be explained just by the twin blessings of *deregulation* and *price rationalization*. Surely, there is a socioeconomic dimension to this change. In essence, deregulation became the lighter for the wood that was ready to be burned. In fact, it should be noted that within Turkey, connection between cities depended mainly on ground transportation by buses and coaches. In 2002, 186 million passengers used ground transportation, with an increase to 236 million in 2010. Thus, the ground transportation market grew by 52 million passengers, whereas the air transportation added only 31 to 32 million new passengers during this period. Thus, the growth of domestic transportation was shared by ground and air. Railroad transportation became insignificant. However, it is interesting to note that total transportation in Turkey, which exceeds 300 kilometers (to exclude intracity transportation) with all modes (ground, rail, air), is forecast to be greater than that of sum of all transportation of the Three Bigs: the United Kingdom, Germany, and France.

We can list at least five structural and socioeconomic reasons during this period of history that resulted in the fast-paced domestic transportation growth:

- The geography of the nation is large (783,562 square kilometers, larger than the state of Texas).
- Migration toward big cities due to work, education, and marriages:
  - İstanbul, as the economic, social, and art capital of Turkey, observes a large influx of people of all ages.
  - Ankara, as the political capital, provides the needs of a large city to inhabitants of Anatolia, thus observes a continuous migration from its hinterlands.
  - İzmir and Bursa, with their moderate climates and various industries, also observe migration.
- Domestic tourism toward a number of regions, especially the Aegean and Mediterranean, increased.
- Gross domestic product per capita increased (from $4,500 in ca. 2002 to $14,000 in ca. 2011), thus increasing the demand for leisure travel.

10.2 Turkish Airlines after Deregulation

In 2003, after deregulation and liberalization of the airline industry, Turkish Airlines initiated its global growth strategy. As a fully state-owned enterprise, Turkish Airlines immediately sensed the vital need to be more competitive after new domestic airlines (Onur Air, Atlas Jet, Pegasus, IzAir) entered this growing market and began capturing the domestic market share.

The management of THY in those years portrayed a vision of global transformation, making THY behave like a private organization in the market. In the midst of 2004, THY grounded a number of inefficient regional jets, reducing aircraft numbers from 59 to 47. Even at that time, a fleet capacity of 100 would not be excess capacity for the growing foreign and domestic markets. THY conducted an interesting strategic analysis by labeling all world cities having 500,000 or more population in the world, and just shade an area of the circle of four hours of flight (supposedly maximum duration of short haul so you can travel back and forth in the same day) to 360 degrees. From this analysis, *İstanbul emerged as the champion world city*, whether maximum number of cities, gross national product covered, or population was scrutinized. The city's location in the world was close to Europe, the Middle East, North Africa, and CIS.[3]

It is a well-established fact that *aircraft utilization* is one of the most critical performance metrics that impacts a firm's profitability. During that time period, a medium-range Turkish Airlines aircraft was typically flown to maximize its use, as the following sample indicates:

- Domestic round-trip flight (e.g., Ankara–İstanbul)
- International round-trip flight (e.g., İstanbul–Rome)
- Domestic round-trip flight (e.g., İstanbul–Trabzon)
- One-way international flight (generally to the Middle East to return at 3 or 4 a.m. the next morning, hence minimizing the idle time)

In such an efficient flight schedule, Middle East passengers can easily be connected to European cities via İstanbul. Using İstanbul as a transfer hub, THY created a competitive advantage among other airlines in the region. For example, in Israel, THY became the second most well-known brand after the national well-established El Al. Iran Air had a difficult time competing with THY on flights to İstanbul. People from the Balkans could easily fly to the Middle East conveniently through İstanbul. THY enjoyed the lack of strong competition from the carriers of neighbor countries. With a combination of these factors, THY succeeded in growing its customer base quickly through international flights. The main drivers for this successful growth plan can be summarized as follows:

- *High use of aircrafts:* Allowed lower prices than its rivals without damaging the service quality.
- *Lower labor costs vis-à-vis European counterparts:* Served competing with Lufthansa, Air France, and British Airways on international routes.
- *Exceptional cabin and premium-quality catering services:* Enhanced the service quality, especially for business-class passengers, increasing the passenger retention rate.
- *Head-on competition with only Emirates and Lufthansa for the Middle East:* Empowered THY to avoid competition with other firms, thereby capturing a larger market share on these routes.
- *International-to-international flights via İstanbul:* Provided flights such as Frankfurt–İstanbul–Singapore, comprising a good chunk of the total flight portfolio. This strategy was instrumental in *hedging demand risk*, as demand for these routes was independent of the domestic and foreign passenger demands bound to and from Turkey.

With such a rapid growth in international markets, İstanbul Atatürk Airport quickly became congested[4] and bottlenecked as expected, a situation that paved the way for the birth of the bright new business model: AnadoluJet.

10.3 Inception of AnadoluJet

In 2007, Turkish Airlines observed five types of passenger traffic in its network:

1. *Local international:* International destination to and from İstanbul, for example, Frankfurt–İstanbul
2. *Connecting international:* From an international destination via İstanbul to another international destination, for example, Frankfurt–İstanbul–Singapore
3. *Local domestic:* Domestic flight to and from İstanbul, for example, İstanbul–Ankara
4. *Connecting international to and from domestic:* From an international destination to a Turkish city via İstanbul, for example, Frankfurt–İstanbul–Adana
5. *Connecting domestic:* From one Turkish city to another Turkish city via İstanbul, for example, Trabzon–İstanbul–Adana

Inevitably, İstanbul had to carry out the first three of these. However, the last two could be well addressed by another airport rather than being bottlenecked at Atatürk. Take the fourth type of passengers presented. These were mainly people of Turkish origin (expatriates) living and working in European countries and visiting different cities in Turkey for leisure purposes (e.g., holiday, vacation, trips to relatives). On those flights, the percentage of business-class passengers was quite low since the majority of passengers were low- to middle-income families. Take the fifth type of traffic. For instance, someone flying from the city of Trabzon in the Black Sea region to the Mediterranean city of Adana does not need to transfer at İstanbul Atatürk Airport. For these two types of passenger traffics, Ankara would be a good potential base.

Ankara, the capital and second-largest city in Turkey, had less than 10% of total domestic flights, implying a great latent potential. A newly constructed airport was attempting to attract carriers that had mostly chosen İstanbul as their home. Ankara was a decent *spoke* for most if not all carriers but none of the airlines had Ankara as a *hub* up until AnadoluJet. *Airline pax statistics (passenger statistics)* so far did not reveal any terms of confidence. However, all Anatolia (i.e., mainland Turkey) needed was Ankara as all governmental offices were based there. Businesses all around Turkey, mainly small and medium-size enterprises, needed to be connected with İstanbul for commercial deals but also with Ankara for governmental deals. Furthermore, as trade among cities in Anatolia[5] accelerated, the need to connect more frequently increased among cities scattered all over Turkey. Apart from housing governmental organizations, Ankara was also home to numerous prestigious universities such as Bilkent, METU, and Hacettepe. It could be easily considered an educational center. Its population topping 5 million created a considerable demand for travel to different cities in Turkey.

Before 2007 a businessman from an Anatolian city who wanted to conduct a deal in Ankara had three transportation options: (1) the very limited number of direct flights to Ankara (i.e., two to three times a week); (2) flights to İstanbul to catch the connecting flight to Ankara (which is a waste of time and money); and (3) ground transportation. In general, people used the third option despite its difficulties. This was the norm although it surely created inefficiencies for conducting business in Ankara. There was an obvious demand that could be satisfied only with daily flights to Ankara from at least twenty cities in Turkey.

Ankara was thus considered a *potential hub*. In terms of its geography, it was located in the center of Turkey. All roads on the ground connecting south to north and west to east had to go through Ankara. More importantly, the new terminal paved the way for fast board-to-board connection, where the minimum time required for connecting flights was drastically decreased to 25 minutes. Moreover, Ankara's centeredness enabled flight times to be around 90 minutes across the board, thus enabling a *bank system* where aircrafts simultaneously depart from and arrive at Ankara around the same time. This synchronic harmony resulted in synergies, including increasing connecting traffic due to disproportionately increased connectivity.

After strategically examining the potential of Ankara as a hub for THY regional flights, a detailed market study was conducted, and in early 2008 passengers of any type who originated from Ankara were studied in detail. The methodology was highly innovative, scrutinizing passengers' mobile phone use. A passenger turning off his or her mobile phone in Ankara and turning it on again after 1–1.5 hours was deemed a typical airline passenger. Another passenger may have turned off the phone for certain duration such as 9–10 hours yet along the way 10–15 minutes of on periods were observed. These passengers traveled via buses and turned off during the trip except for the stopovers on the route. Last, some passengers traveling via trains or private vehicles did not need to turn off their mobile phones, and their routes could be identified. The study's findings revealed that one of every seventeen passengers used air travel from Ankara. Surveys inquired as to why the remaining sixteen people did not fly: The main reason was price. If the prices were lower, they would prefer flying one hour instead of traveling on the road for seven to eight hours or more. Also, these passengers did show a reluctance to pay extra for these short-haul flights for the amenities offered by THY such as CIP (commercially important passengers) lounges, business class, and sophisticated catering. Even more interestingly, more than half of the people in the survey said that had it been possible, they would even *fly standing up* in the aircraft as long as the price was going to be reduced by half.

Considering all of these factors, THY decided to fulfill the unmet need for a low-cost product and hence to isolate itself from differentiated standard services offered by the mainstream brands. It initiated a new brand, called AnadoluJet, literally meaning "Anatolia Jet," with a vision of connecting Anatolia with virtual flight bridges. Ankara, the capital and the geographical center of Turkey, served as the main hub of AnadoluJet. The brand, as a subsidiary of THY, started its operations in April 2008. Despite not being a separate entity of THY, AnadoluJet was independent from mainline business in its entire strategic, commercial, and operational decisions, such as pricing, scheduling, network planning, marketing, advertising, and catering.

10.4 Operational Innovations in a Risk Intelligent Supply Chain

AnadoluJet excels at many dimensions of operations in its risk intelligent supply chain. It is a unique case for operational innovations.[6] As opposed to world-known, low-cost carriers that fly point to point such as Southwest Airlines, JetBlue, and Europe's Ryanair and easyJet, AnadoluJet uses a *hybrid model*. This model uses the best-in-class features of both the *low-cost carrier (LCC)* and the *legacy carrier strategies*. In the LCC strategy, a flight hub is created just by coincidence. Increasing aircraft use is the key goal without using flight transfers. This strategy was deemed inappropriate for AnadoluJet for reasons explained in the following sections. Thus, AnadoluJet created its own unique business model in the market to become the fastest growing airline brand in Turkey. Growth rates were phenomenal: during 2008–2009, it reached 62% and during 2009–2010 it hit 80%.

10.4.1 Flight Schedule Orchestration

By deductive reasoning, AnadoluJet first wanted to clearly understand the passengers' preferences in terms of flight schedules in Turkey. AnadoluJet does not offer any early morning flights (i.e., between 5:00 and 6:30 a.m.). Flight time should be reasonable for passengers, so morning starting flights were scheduled at around 7:00 a.m. from different cities in Turkey flying toward Ankara. With 8:00 to 8:30 a.m. arrivals, passengers could conduct business during the day and toward the evening could easily fly back to their originating cities throughout Turkey.

A flight network was deliberately designed to connect as many cities as possible to each other in Turkey. In such a network, Ankara is the *connecting or transfer hub* for a passenger who flies from city A to B, for example, from Gaziantep to Trabzon.[7] Dedicating Ankara as a transfer hub borrowed from the *hub-and-spoke system* of legacy carriers. In such a strategy, transfer times should be short and the flights should be carefully synchronized. This is what AnadoluJet achieves with great success. Since Ankara is strategically located in Turkey, within 45- to 75-minute flight times of any city, keeping transfer times to 35 minutes on average enables a passenger to reach his or her final destination within 3 to 3.5 hours at most. Such a service is highly desirable if it can be priced correctly.

To achieve this level of flight times, flight schedules need to be *highly homogeneous, synchronized*, and *modular*.[8] Each aircraft makes eight trips per day beginning in a city in Turkey, flying toward Ankara in the morning, and then completing seven more trips until its final destination again not at the hub, Ankara, but an Anatolian city, hence accomplishing four connecting banks to the hub rather than three when compared with the case where the first flight had initiated from Ankara. This highly synchronized and well-orchestrated operation echoes Zara's fixed delivery schedules in the fast fashion industry.[9]

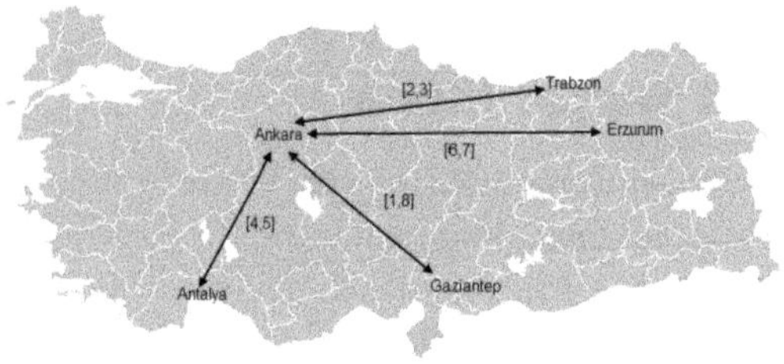

Figure 10.1 An example of a typical domestic daily flight schedule: Beginning and ending at the city of Gaziantep, while flying to and from three more cities (Trabzon, Antalya, and Erzurum).

Here is an aircraft's daily flight schedule in detail, also displayed in Figure 10.1.

1. Gaziantep–Ankara (around 7:00 a.m.)
2. Ankara–Trabzon (around 9:00–9:15 a.m.)
3. Trabzon–Ankara
4. Ankara–Antalya
5. Antalya–Ankara
6. Ankara–Erzurum
7. Erzurum–Ankara
8. Ankara–Gaziantep (around 10:45 p.m.)

An aircraft flies to Ankara and then visits three more cities before returning to its origin. In this strategy, the aircraft spends the night in its origin city. It is clear that an LCC strategy would never allow an aircraft to be grounded for the entire night. Considering that AnadoluJet pays for the accommodation and ground transfer of the crew in these cities, this strategic choice results in extra costs. Nevertheless, using genuine pricing and managing the fleet intelligently, AnadoluJet creates more value to compensate for these extra cost items. In this flight schedule, we need to note that the first flight from Ankara Esenboğa Airport to another city in Turkey is around 9:00–9:15 a.m. This time schedule is also deliberate, considering the thirty-minute drive to the airport from downtown Ankara. Using Ankara Esenboğa Airport as a hub considerably reduced the domestic transfer load of İstanbul Atatürk Airport, shifting it toward Ankara. Right now, 80% of the domestic flight transfers are operated by AnadoluJet. See Figure 10.2 for the flight map of AnadoluJet (as of August 2012).

10.4.2 Genuine Pricing Strategy

It is clear that without the *demand depth* created by Ankara, it would be very difficult to justify the costs of low aircraft use, that is, grounding the aircraft for the whole night in various Turkish cities. Also, the passengers would not prefer AnadoluJet

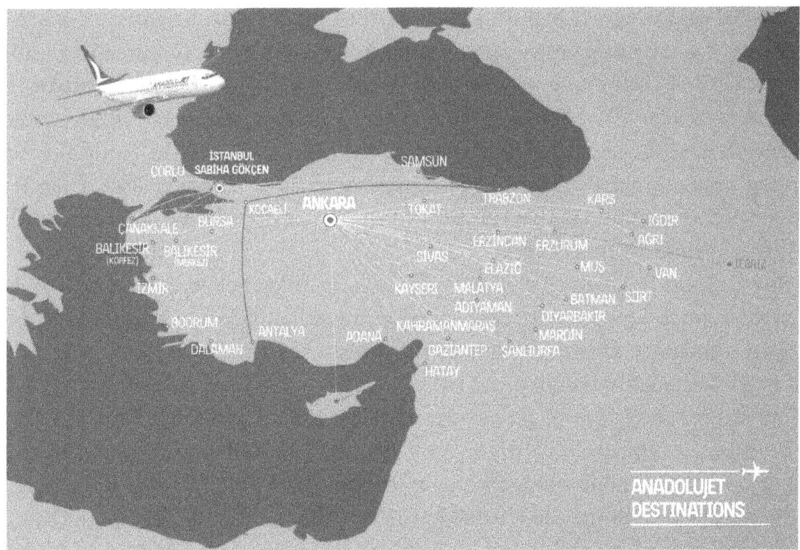

Figure 10.2 AnadoluJet flight map in August 2012. (Courtesy of AnadoluJet.)

unless its ticket price converges with ground transportation cost (i.e., bus, coach). Demand is thus triggered primarily by prices. And price sensitivity is quite high.

While determining pricing strategy, ground transportation prices were analyzed in detail. For the Ankara–İstanbul trip, bus tickets were sold at 25–30 TL. Likewise, a trip between Ankara and İzmir cost 40 TL, and for the farthest trips, such as between Ankara and Van, bus tickets were sold around 75 TL one way. For a passenger to choose air travel, the ticket price could not be more than 20–50 TL more than what he or she paid in bus fare. Therefore, on average, an 80 TL ticket price was considered a good start. However, further cost reductions were put into practice through other operational innovations.

First, the capacity of the aircraft was increased by 15% by reducing the legroom to 29 inches. This space is adequate for a person of maximum 6 feet 3 inches (190 cm) in height. However, if the person is taller, AnadoluJet reserves eighteen different seats in the cabin such as the ones at the exits and the first-row seats to accommodate such a need. Hence, the Turkish national basketball team 12 Giant Men, consisting of twelve players, comfortably flew with AnadoluJet.

Second, without eliminating food and drink service onboard, some parts of the galley (food section per se) were completely removed to help seating capacity and weight reduction. Passengers are provided the choice of complimentary tea and coffee, water, sandwich, and muffins presented in a basket instead of on trays and plates. Also, for eight daily flights, loading the food is done every four flights, reducing the high handling costs. Additionally, in 737-700 aircraft, two instead of three restrooms were deemed sufficient. These enabled an increased number of available seats on the same aircraft and thus reduced the cost per seat. AnadoluJet is keen on upgrading the passenger experience. Since passengers are accustomed to complimentary tea and coffee and muffins during bus trips, removing such onboard treats would downgrade the passenger experience and therefore was strictly avoided. Furthermore, selling hot food onboard like the other LCCs of the world do creates extra logistics costs that may not be compensated for on many routes. Even if it is profitable for a limited number of routes, the lack of standardization across routes significantly hinders this process.

Third, AnadoluJet does not provide loyalty miles below a level where THY prices start. Such low prices do not bring mileage liability. More than half of sales are done via direct channels such as the Internet, call centers, and sales offices, resulting in reduced sales channel cost.

Fourth, the *load factor* is on average 85% compared with 70% for THY flights. For example, a typical THY flight with Boeing 737-800 would have 100 to 102 passengers of 159 available seats, whereas AnadoluJet would fly with 160 passengers of 189 seats. Taking into consideration the fact that passenger-dependent costs are totally negligible when compared with flight-dependent costs, 160 passengers who each paid 100 TL is equivalent to 100 paying 160 TL. Once the price was lowered from 160 TL to 100 TL, it is not surprising to find 160 people instead of 100 on the same flight. The combination of *high price sensitivity* with *market depth* was the key success factor.

Last but not least, AnadoluJet greatly emphasizes genuine pricing campaigns. In the winter period, it is the only airline that offers *guaranteed advance purchase* tickets to the market such that a passenger could buy any domestic route ticket at 59 TL (approximately US$30 including taxes) if he or she buys at least seven days in advance. The only exception is Fridays and Sundays, where AnadoluJet does not provide a guaranteed low price. Market demand determines price offerings. "Unlike other airline campaigns that typically says "starting from x TL," AnadoluJet has changed the rules of the game," says Sami Alan, senior vice president of regional flights for AnadoluJet. "Customers just don't like campaigns that outsmart them with *fine print* exceptions," (private communication with Sami Alan, Spring 2012) he adds.

AnadoluJet designs its pricing campaigns to be plain, simple, and comprehensive. It wants the passengers to understand the process and be persuaded to purchase. "Thanks to Internet technologies, prices are more transparent than ever and hence power has shifted to the customer. This is why advance purchase campaigns that are communicated well are pretty successful," (Ibid.) says Alan. Once outdoor advertisements showed up with "Book three days in advance, all Tuesdays all seats everywhere 59 TL," the number of employees at the call center had to be increased.

10.4.3 Fast-Growth Network Management

Since the growth rates are 62% (2009) and 80% (2010), the brand needs to be managed in a fast-paced environment. When the brand was developed, aligned with its vision of connecting Anatolia, it positioned itself in a niche market, providing reasonably low prices without lowering the basic essential standards like on-time performance, for which THY is known. AnadoluJet capitalizes on the trust already built into the prestigious THY brand. Since the name directly resonates with the people who live in Anatolia, the market fully embraces it. Moreover, passengers perceive that they are obtaining smart deals together with the THY service quality since AnadoluJet provides a low price of 59 TL one way and a transfer price of only 45 TL for the connecting flight, all taxes and surcharges included.

AnadoluJet knows that it is imperative to increase the connectivity index of the cities in its network. The connectivity index refers to how well a city is connected to every other city in the network. This index also demonstrates the connectivity depth of the cities. The higher this index, the more flight options passengers have throughout one day. Growing the flight network by adding new cities implies a disproportionate increase in the connectivity of the cities. For instance, a growth of 7% in the number of cities in the network leads to an almost 700% increase in the connectivity because the relationship is exponential.

AnadoluJet aims to be the first brand to be known and requested by potential passengers in Turkey. As Alan puts it succinctly, "As you become the first address to be known, you start winning in the market." Building on this goal, as the scale of the operations increased, AnadoluJet observed an increasing return of passengers and has become the first brand to be requested at sales points in most parts of Turkey. With forty domestic destinations, AnadoluJet by far has the maximum number of cities covered in Turkey, as the other airlines are around or below twenty. Besides, from major cities in Turkey (e.g., İzmir, Antalya, Adana, Trabzon, Erzurum, Bodrum), AnadoluJet is the only carrier to fly to both İstanbul and Ankara.

On its path of continuous growth, AnadoluJet chose Sabiha Gökçen Airport, located on the Asian side of İstanbul, as its secondary hub after Ankara Esenboğa Airport. This move has been advertised using the motto, "We are full force at Sabiha Gökçen!"[10] AnadoluJet started its Sabiha Gökçen operations with only two aircraft; as of now it flies five, but in summer 2011 it began operating seven aircraft at this airport. AnadoluJet is fully aware of the fact that its competitor Pegasus Airlines uses Sabiha Gökçen as its main hub. As opposed to Pegasus, which offers frequency depth to and from Sabiha Gökçen Airport, AnadoluJet aims to reinforce its strategic position created at Ankara Esenboğa Airport hub by adding new routes from Sabiha Gökçen. This strategic move was deemed necessary to succeed in becoming the best-known address in the industry. Thus, the goal is achieved by adding Sabiha Gökçen flights to its network. For example, a passenger flying from Adana to different cities in Turkey, including İstanbul, can achieve this by using AnadoluJet and connecting within one hour via the Ankara hub. However, this Adana passenger also wants to fly directly to İstanbul instead of transferring via Ankara. This necessity brought the strategy of the *all-in-one* service offering. In this offering, by adding one more flight from İstanbul Sabiha Gökçen Airport to the basket of already-existing flights via Ankara, AnadoluJet is able to increase the market coverage and satisfy its cost-conscious customer base considerably.[11] To further strengthen this position, AnadoluJet chose a decent basket of niche destinations such as Çanakkale, Sivas, and Iğdır and capitalizes on the fact that AnadoluJet flies to everywhere in "Anadolu" as directly as possible.

To manage this fast-growing network, AnadoluJet ought to have *flawless operations*. Such a goal is one of its management's top priorities. AnadoluJet boasts a transfer rate of 99.8% at the Ankara Esenboğa Airport. The rest of the flawed transfers, being 0.2%, are deemed very important. Those passengers who experience delays are transferred to premium hotels and pampered at the highest level possible. In no case does AnadoluJet provide an excuse for a flawed transfer process. It intimately knows the cost of bad word-of-mouth. To be able to achieve thirty-five- to forty-five-minute turnaround times at Ankara Esenboğa Airport, on-time departures should be higher than 90%. Actually, recent data show that on-time departures fluctuate around 94% to 95%, which is best-in-class achievement in the industry. To sustain this level of punctuality, AnadoluJet employs *wet-lease backup*—aircraft and crew leased from other airlines but painted in the scheme of the operating carrier—in each hub during the winter season when weather causes the vast majority of operational disruptions. "Connectivity sits at the core of our model. Charter carriers have excess capacity in winter, and it is fairly affordable to have idle backup support from charter carriers this time. It is a win–win case for all parties: passengers, charters, and us" (private communication with Sami Alan, Spring 2012), says Alan.

Fast turnaround times are achieved through a few key operational innovations. These innovations reduce the boarding time substantially and mitigate flight disruption risk. First, AnadoluJet completely eliminated the x-ray security checks at Ankara Esenboğa Airport for transfer passengers flying into the airport from another city. Since these passengers are already security checked at their originating cities, a second security check at the transfer hub is clearly redundant. However, this operational innovation required regulatory approval. Therefore, AnadoluJet prepared a procedure for approval by the City Security Council and the Mayor's office. Second, delays may occur when a passenger does not have an ID card ready at the check-in process. Knowing that this undesirable event can cost time and potentially disrupt the flight, AnadoluJet created another operational innovation. A police commissioner stationed at the airport can now provide a temporary ID card valid for only a single flight so that the passenger can be admitted without disrupting the whole flight. These two operational innovations reduced the boarding time of transfer passengers, thus yielding thirty-five- to forty-five- minute turnaround times.

AnadoluJet flies to a number of cities that no other rival airline flies. This strategy caters to *fringe customers*[12] who have not been provided any flight services previously. Interestingly, once people become aware of such services at low prices, they fill up the flights long before the departure times to reduce their chance of not finding seats and going through tedious and long bus trips.

Today, AnadoluJet's number of city pairs—cities that can be connected by air—is greater than twice those of its rivals (Pegasus, Atlas Jet, and SunExpress), which creates a variety of flight options to potential passengers.

10.4.4 Intelligent Fleet Management

AnadoluJet leases and operates the newest aircraft.[13] All its aircrafts were constructed in 2007 or later. As of now, AnadoluJet operates Boeing 737-700 and Boeing 737-800 models in its fleet. AnadoluJet steadfastly believes in *fleet commonality* for various reasons. However, having a common fleet need not mean operating one type of aircraft. AnadoluJet prefers *one family* in its fleet as opposed to having *one type of aircraft,* unlike its competitors Pegasus and SunExpress, that operates solely one type of aircraft (i.e., Boeing 737-800s). In its total fleet of 22 aircraft, 10 offer 149 seats, 6 offer 177 seats, 2 offer 180 seats, and the remaining 4 have 189 seats. Aircraft with 177 seats, leased from THY, were transformed to 189-seat aircraft. The rest of the fleet were leased from the market. AnadoluJet constantly uses its peripheral vision[14] to deliberately search for new opportunities. For example, when SkyEurope went bankrupt, AnadoluJet leased eight of its aircraft, all Boeing 737-700s, at bargain prices. Leasing one- to two-year-old aircraft as opposed to brand-new ones reduced the leasing costs by 15% to 20% and also substantially lowered the repair and maintenance costs.

Fleet commonality and a one-family aircraft strategy offer several benefits, the first of which is disruption risk mitigation. In a flight-delay scenario, crew can be transferred to aircraft with the same features without disrupting the operations. The second benefit is lower fuel costs, fewer cabin crew members, and lower leasing costs. AnadoluJet's smaller-capacity (149 seats) aircraft fly new routes. The cost of such a single trip on a new route is 15% lower than the trip with a 189-seat capacity aircraft. Using this smart strategy, AnadoluJet can provide four flights to a recently opened route instead of three flights, increasing the flight frequency and thus offering more options to passengers. This helps in capturing a greater share of the potential demand.

A simple illustrative example is presented in Table 10.1. In the example, offering 5% more flight capacity by increasing the total costs only by 13.3%, yet achieving this in four flights instead of three, is a strategic choice made by AnadoluJet to capture more of the market; hence, it bolsters the goal of becoming the best-known address. Third,

Table 10.1 Simple Illustration of Route Opener Strategy: Capacity versus Flight Costs

	149-Seat Capacity Aircraft	*189-Seat Capacity Aircraft*
Total Capacity Offered	4 flights × 149 = 596 seats	3 flights × 189 = 567 seats
Cost of the Flight	4 × 85 TL = 340 TL	3 × 100 TL = 300 TL

AnadoluJet also leverages upward capacity flexibility in its operations. If a certain city in its network has a special event such as a festival, sports game, or concert for a certain period of time, AnadoluJet can quickly improvise and switch to a larger-capacity (189-seat) from a smaller-capacity (149-seat) aircraft to capture the ephemeral demand spike within the same family of aircraft. In sum, fleet commonality with a one-family aircraft strategy provides AnadoluJet both cost and operational flexibility, while mitigating demand, capacity, and disruption risks. With all these strategic choices, AnadoluJet has been well equipped for intelligent risk taking in its supply chain network.

10.4.5 Smart Revenue Management

AnadoluJet firmly believes that dynamic capacity management should be aligned with dynamic pricing. The company makes smart use of analyst intuition on top of classical revenue management techniques. First, it carefully analyzes the domestic and international flight demand patterns. International flights are generally booked outside of three days before the flight date, whereas domestic flights do not sell well more than seven days before the flight. Furthermore, tickets are sold much more quickly in the last seven days for domestic flights. To handle this minutely fluctuating demand, AnadoluJet dedicates twenty revenue management analysts with exceptional math skills. They work in an atmosphere similar to that in which a stock market analyst works, observing the demand changes in front of their dashboards. Every revenue analyst currently focuses on five routes in an average per day: approximately twenty-five to thirty flights. Using intelligent warning alerts when certain demand levels are attained for given price levels, analysts make the necessary changes to the seat prices. Mainly, they open up strategic slack capacity close to the flight dates. Timetabling is for twenty-two airplanes, two being the strategic capacity (also being an operational hedge if necessary.) Based on demand pattern, AnadoluJet could offer two flights with larger-capacity aircraft or three flights with reasonable prices that would sweep the whole market. This flexibility enables AnadoluJet to be more nimble and responsive in such a fickle market.

AnadoluJet overbooks its flights to minimize the risk of empty seats. Yet overbooking levels are miniscule, such as two to three seats per flight. Usually its average overbooking is one standard deviation away from the mean; for example, a flight with a 189-seat capacity can be overbooked for 191 passengers.

10.5 No-Excuse Customer Service

AnadoluJet knows that investment in passengers is the ultimate value-generating strategy, although in the airline industry, passenger satisfaction is not an easy task. A successful flight operation is considered the norm. Slight deviations from it can upset and aggravate passengers instantly. Therefore, AnadoluJet aims to understand its passengers intimately by satisfying their needs on time. The business model in the *customer service team (CST)* works as follows.

Passenger complaints arrive at the CST. A response is given generally within four days. If the complaint cannot be satisfactorily resolved during that period, then on the fifth day it is sent up to the chief, then to the manager, and technically all the way to Alan, who then takes the stage and personally responds to the passenger. If a passenger is really upset, AnadoluJet provides not only a formal apology but also a compensation package such as VIP transfers in the city, a complimentary ticket, and hotel awards. Interestingly enough, AnadoluJet also receives approximately 10 apologies daily from passengers who might have overreacted when interacting with CST. For these passengers, AnadoluJet likes to surprise them with free tickets. Alan summarizes their approach succinctly: "In this business, what counts is what you do, not what you say or what is being said about you." So AnadoluJet practices what it preaches by taking firm, fast, and proactive actions toward its passengers to strengthen its ongoing relationships.

10.6 Designing Responsive Marketing and Sales Strategy

According to AnadoluJet, a *responsive marketing strategy* is the key in the airline industry. The marketing statement and how it is being communicated play a pivotal role as the power has shifted to the passenger. AnadoluJet uses a multitude of channels for communicating its value proposition for a customized and effective message. For example, in Cyprus, due to the low population density and a wide geographic area, local radio channels are used for advertisement. On the other hand, in Turkish cities, outdoor advertisements are installed to reach a greater population. For hotels, AnadoluJet has developed two million door hangers, which state, "We are not home since we are flying with AnadoluJet." For some towns in Turkey, such as Çorlu and Tekirdağ, marketing function has been conducted by a team of four people through *one-on-one marketing*. The team has visited all of the households in one year and informed the residents about the AnadoluJet flights coming to town.

Unlike competitors posting advertisements stating that prices start at 19 TL yet can reach extremely high levels, AnadoluJet is keen on providing a clear and simple message to its potential passengers. It is critical that the message should persuade passengers that ticket prices cannot be extremely high. Thus, for a domestic ticket, the price begins at 59 TL and is capped at 199 TL. AnadoluJet sells tickets mainly through four sales channels with the following proportions:

- Direct sales online (30%)
- Direct sales via the call center (10%–12%)
- AnadoluJet sales offices (8%–9%)
- Independent sales agents (50%)

AnadoluJet designed a website (http://www.anadolujet.com) that is simple, user-friendly, and reminiscent of Google. In fact, Alan's goal is that this website eventually will be accessible to the approximately 10% of Turkey's population that is unable to read or write. Thus, achieving a simpler design is everyday work for designers. When a user searches for tickets on the website, the two cheapest fares—the promotional fare and the flexible class—are presented to the passengers with the options of plus or minus three calendar days. AnadoluJet invests in call center management as passengers are more willing to purchase with real sales agents. The rest of the ticket sales are made via sales offices and independent agents. In Anatolia, sales agents are still the most trusted entities for ticket purchasing. These agents provide a safe harbor in case of problems during the process, and they also operate under longer payment terms (sometimes on credit), which are certainly more favorable to most passengers. Hence, 50% of the passenger population leads toward the independent sales agents.

To further enhance ticket sales, AnadoluJet created an innovative campaign called Jetgenç (http://www.jetgenc.net) that caters to university students twenty-five years or younger. In this campaign, first a student registers with his or her name and mobile phone number. When other individuals purchase tickets online, if they provide the student's mobile phone number, the student earns one free ticket once ten tickets are sold. Through this marketing campaign, AnadoluJet essentially motivates students to become its sales agents. In one year, the campaign became a success.

Another innovation was accomplished in *service bundling*: "Fly with AnadoluJet and land in front of your house" is the motto of the partnership with *SecureDrive,* which provides transfers for passengers from and to the airports. Employing a big fleet of AnadoluJet-branded vehicles, SecureDrive sells transfers within İstanbul, Ankara, İzmir, Antalya, Bodrum, and Dalaman—with Trabzon, Adana, Diyarbakır, and Northern Cyprus to follow. Customers enjoy a 30% discount for already low prices. A typical transfer fee from anywhere in the city of Ankara to the airport is no more than 30 TL, which is less than half the cab rate. AnadoluJet attains not only increased customer service but also a good 10% of the revenue generated.

A project currently implemented regards parking privileges to AnadoluJet customers. AnadoluJet *Park & Fly* resorts will be located next to airports, and AnadoluJet customers will enjoy first-day-free parking and the other days only 10 TL/day. This will not only create a competitive advantage for AnadoluJet but also will drop a few million TL all the way to the bottom line of the income statement. "Ancillary revenue creation is the buzz nowadays in the LCC world; without creating additional value to the customer we just can't justify any additional income. Yet there are many ways to serve and earn more," summarizes Alan.

10.7 Developing Young and Ambitious Talent

Alan, a high school mathematics Olympiad bronze medalist, is a firm believer in *competence* and *capability*.[15] Experience comes as a second priority when selecting his team members. Right now, 100 young and bright people work for the AnadoluJet brand. Additionally, there are 40 pilots and 300 cabin crew members, most of whom are based in Ankara. Additional crew members are called on duty as needed from İstanbul. With a starting base at Ankara Esenboğa Airport, AnadoluJet had no other option but to develop and train the talent himself from scratch. What is striking is that now almost two-thirds of the management team is twenty-five years or younger. Moreover, 70% to 80% of the team ranked in the top 1,000 among 1 million university candidates in nationwide aptitude tests. In short, they are the gifted, results-oriented, and hard-working youngsters who would benefit from mentoring. Alan knows this and therefore constantly hunts for the following qualities while selecting his management team: (1) capacity and capability to learn; (2) ambition to succeed; and (3) team spirit with modest character.

Alan prefers promoting young talent with open and sharp minds rather than converting experienced and possibly sluggish mind-sets. For example, one of his teammates is a twenty-five-year-old industrial engineer who is managing a team of six and is fully responsible for the flight schedules and the aircraft contracts. A typical analyst, in his or her early twenties, is managing a US$50 million business portfolio. Having such young talent in managerial positions provides AnadoluJet with a unique opportunity to use the very different perspectives of the millennial generation.[16]

For the cabin crew, AnadoluJet hires mainly graduates of prestigious secondary schools, such as TED and Yükseliş Koleji in Ankara, and pays them competitive salaries. These graduates have a good command of English, effective communication skills, and high emotional intelligence. Since the passenger experience occurs in the cabin, the quality of the crew is of utmost importance.

10.8 Risk Management Philosophy in Orchestrated Network

AnadoluJet's main philosophy for risk management is based on *outrunning the bear*. As long as the company is better than the competition, they can outrun that bear and be ready for potential risk events in Swan Lake.

Fuel price, one of the most critical and volatile cost factors in the global airline industry, is determined by market forces. An increase in the fuel price reflects a general trend of global market growth. This growth is expected to lead to more demand for air travel. Thus, AnadoluJet bets on the self-stabilization effect of fuel price increases. On the volatility side of the fuel prices, fuel price hedging is not seen to be the right tool by AnadoluJet. AnadoluJet thinks that in the long run, one cannot outsmart the market forces of hedge financiers. However, the parent firm THY dedicated a small team of ten to fifteen people to learning the intricate details of fuel price hedging. The chairman of Pegasus Airlines, Ali Sabancı, stated that in 2008 a bad experience with fuel hedging cost the firm 5.1 million Euros.[17]

Instead of risk hedging with financial derivatives, AnadoluJet is keen on managing strong financial security for unforeseen events (black swans) in the future. For catastrophic events such as the recent volcanic ash mishap in Iceland, AnadoluJet believes that lessors would not prefer that airlines go insolvent.[18] Thus, having a *strong liquid position* is crucial in such situations until the whole industry begins taking off again.

On the other hand, AnadoluJet diligently uses operational hedging strategies for risk management. For instance, *two backup aircraft* are ready in winter season waiting with their crews at the Ankara Esenboğa Airport for unexpected events (cancellations, disruptions due to external weather risks such as flooding and snowstorm, and natural disasters like earthquakes) in the flight network. AnadoluJet does not tolerate more than three-hour delays. In such cases of inevitable delays, flights have to be cancelled, which should be strictly avoided if possible.

As another effective operational hedge, AnadoluJet negotiates for *flexible aircraft leasing contracts.* AnadoluJet is attentive and closely monitoring the changes in global aircraft design and manufacturing. Engine and material innovations by Boeing and Airbus, as well as the moves of Japanese, Russian, Chinese, and Brazilian firms, are expected to increase industry competition. These engine and material innovations are expected to lower the price–quality ratios. Hence, AnadoluJet pushes for contracts with more favorable terms. In sum, AnadoluJet keeps *flexible timing options* in its contract portfolio to buy or lease future aircraft and also tries to negotiate better deals when possible with the lessors.

10.9 The Path Forward

The future for AnadoluJet looks bright. Managing a growing domestic and international flight network is not an easy undertaking. AnadoluJet's awareness of this fact is reflected in its ambition, innovativeness, and humbleness in the industry. Avoiding head-on competition with its rivals and being risk intelligent with its strategic and operational moves, AnadoluJet is successfully orchestrating its flight concertos as a maestro over the clouds. And it seems that the concert over the clouds will not finish soon.

10.10 The I-Quartet Model of AnadoluJet

The I-Quartet model of AnadoluJet's RISC is summarized in Table 10.2. It is easy to see that the four roles of the I-Quartet model work together for AnadoluJet. However, the roles of Inquirer, Improviser, and Ingenious are slightly more visible in AnadoluJet's risk intelligent supply chain management. It is true that with a new business model, it is necessary to constantly learn with a peripheral vision from failures, near misses, and weak signals. This is achieved in the Inquirer role.

Table 10.2 AnadoluJet's I-Quartet Model

Integrator	*Inquirer*	*Improviser*	*Ingenious*

• Fixed, homogeneous, synchronized, orchestrated and modular flight schedules
• Customized but integrated advertising strategy
• Orchestrated no-excuse customer relationship management
	• Honed skills and capabilities at Ankara hub, transferred to secondary hub Sabiha Gökçen Airport
	• Full authority on ground operations given to AnadoluJet management for Ankara and Sabiha Gökçen airport
	• Constant learning and improvement in operational performance metrics
	  ◦ Passenger transfer rates via Ankara Esenboga Airport are 99.5%
	  ◦ Average turnaround times are 35 minutes
	  ◦ On-time-departure rate is 94%–95%
	• Using intelligent warning signals in revenue management and pricing
	• Deliberate opportunity seeking in the global aircraft market with peripheral vision
		• Operational improvisations for agility and speed
		  ◦ No second security checks at Ankara hub transfers
		  ◦ Temporary ID card processing
		• Fleet commonality
		  ◦ Firm belief in aircraft family, not aircraft type
		  ◦ Fleet capacity flexibility
		    ▪ 149- and 189-seat Boeing 737
		    ▪ Downward flexibility (route opener, flight frequency depth, low risk)
		    ▪ Upward flexibility (capture ephemeral demand spikes
			• Operational risk hedging
			  ◦ Two backup aircraft ready with crew for unforeseen contingencies
			  ◦ Wet-lease backups
			  ◦ Flexible aircraft leasing contracts
			• Holding strong cash reserves
			• Risk management philosophy: Outrun the bear
			• Catering to fringe passengers on new routes that no other airline flies
			• Using strategic slack capacity in revenue management

AnadoluJet has to create variable responses to unexpected disturbances along the way. This process requires improvisation, adaptation, and creative tinkering, which are handled in the Improviser role. The Ingenious role forms the backbone of intelligent risk taking in the supply chain network. Last, AnadoluJet uses the Integrator role to orchestrate its flight as well as supply chain networks to ensure resilience and robustness.

Epilogue: Where Is Our Next Journey?

> We must always change, renew, rejuvenate ourselves; otherwise we harden.
>
> **—Johann von Goethe**[1]

> Live your life each day as you would climb a mountain. An occasional glance toward the summit keeps the goal in mind, but many beautiful scenes are to be observed from each new vantage point. Climb slowly, steadily, enjoying each passing moment; and the view from the summit will serve as a fitting climax for the journey.
>
> **—Harold B. Melchart**[2]

> Living is a form of not being sure, not knowing what next or how. The moment you know how, you begin to die a little. The artist never entirely knows. We guess. We may be wrong, but we take leap after leap in the dark.
>
> **—Agnes de Mille**[3]

If you have arrived at this part of the book, you have probably realized by now how today's global supply chain networks operate in a *Swan Lake* of multiple risks—ranging from white to black swans. You have also seen that a risk intelligent supply chain operates effectively with four roles of the *I-Quartet model*: *Integrator, Inquirer, Improviser,* and *Ingenious*. A risk intelligent supply chain is always in flux, continuously at the edge of order and chaos.

After appreciating the essential qualities of a risk intelligent supply chain, you have also enjoyed conversational journeys with a number of key Turkish decision makers who manage these risk intelligent supply chains day in and day out. In this section, I want to emphasize seven key strategies for orchestrating and managing a risk intelligent supply chain.

1. Do not just focus on the black swans. Manage the Swan Lake habitat. There are white and gray swans of diverse sizes and shapes craving for your attention.
2. In Swan Lake, protect your firm against negative black swans, but be open and flexible, sometimes even intelligently aggressive, toward positive black swans. Continuously manage the balance of coupling strength and heterogeneity in your suppliers, customers, partners, and employees so that positive black swans are possible.
3. Work on designing a supply chain network with a peripheral vision that senses and learns from failures, successes, near misses, precursors, and weak signals. In this process, always be aware of your and other stakeholders' perception biases.
4. Not only anticipate early warning signals in your supply chain network, but also interpret and act upon them. Employ strategies such as the *Shivers model* to anticipate supply chain sickness.
5. Know that you are part of a greater and wider supply chain network over the globe. Work toward enhancing the resilience of this supply chain network through improvisation using creative tinkering and design thinking. Learn through the supply chain network wisdom by ensuring the right level of diversity and flexibility.
6. The weakest nodes/links in the supply chain network can move dynamically and unexpectedly, as does a moving knot of a *Turkish Carpet* in the weaving process. Continuous improvisation is necessary to orchestrate a supply chain network. You need to be highly adaptive and responsive to changes inside as well as outside your supply chain network.
7. Know that the impact of risks in a supply chain network can be interdependent and cascading. Incorporate the risk interdependence of supply chain network stakeholders into your decision making.

I want to finish the book with one last metaphor, which I call *Ocean Exploration.* Until now, we have seen various features of the risk intelligent supply chains in a multitude of settings for different companies. Now, I hold the mirror to you, at an individual level, because the presence of each one of you is essential in creating and managing your firm's sustainable risk intelligent supply chains.

Perceptions, mind-sets, actions, preferences, and strategic perspectives of the Turkish executives in this book are hopefully now in your so-called ocean of risk insights. You can dive into this ocean whenever you wish and explore its amazing and interesting creatures. Some of these inhabitants will truly surprise you, yet others will be familiar and perhaps not very interesting. Please continue this exploration without any preconceptions and higher expectations. One day, you will hopefully hit the blockbuster moment such as seeing the great white shark from a short distance or swimming with a *caretta caretta* (loggerhead sea turtle).

Diving into the ocean was taking the risk in the first place, yet you have been preparing for this moment all your life. You have learned the elaborate details of the ocean and how it functions with its inhabitants. Once you are in the ocean, enjoy the scenery, and always open your eyes, being mindful about those unexpected, joyful moments. During the journey, always update your beliefs and perceptions, learning constantly. In the business world, each company and supply chain network is swimming in the ocean of risk insights, sometimes without being aware of it. Hence, the cost of mindlessness is immense.

Exploration by diving into the ocean is what we perform in this age of fragility, uncertainty, and turbulence. This age and the following ones seem to disrupt our lives in unforeseen and unexpected ways as complexity and interdependence increase—just as discussed in this book. These disruptions, like ocean currents, paradoxically enough, carry the seeds of making our lives happier, peaceful, and more content as oceans carry seeds of various sorts from one continent to the other.[4] This age is also full of opportunities that will bring the booms, breakthroughs, and blockbusters into our lives. We have all witnessed the Internet's rise and presence in our lives as well as the rise of mobile technologies.

All in all, fragility and resilience are two sides of the same coin. They go hand in hand in any event. Unless you learn and accept your fragility at a personal level, you will never be resilient in times of stress and turbulence. Hence, the most fragile are the most resilient. Consider ants and bees as well as butterflies. They had been living on earth for millions of years when the dinosaurs became extinct.[5] Water is the softest and most fragile, yet if it drips continuously and patiently, it penetrates marble. So, let us persevere, learn, improvise, and wait for the great surprises in our lives by being aware of our fragility as well as our resilience.

Appendix A: Scholarly Readings on Supply Chain Risk

Supply chain risk management literature in the context of supply chain management is like an ocean. In this section, I provide a number of important pointers to some of the most popular articles that became pioneers in the field. Thus, the brief review is not comprehensive. Surely, as is always the case, a good scientist creates more questions than answers. So my aim is to create more questions and encourage further study in this area by placing signposts on your path.

A number of books have come out in the past few years that address supply chain risk management both academically and practically. Brindley (2004), Zsidisin and Ritchie (2010), and Khan and Zsidisin (2011) are by-products of the International Supply Chain Risk Management Network (ISCRIM).[1] They cover various aspects of supply chain risk management, providing global case examples in different industries as well as conceptual models. Mathematical modeling-oriented people can consult Kouvelis et al. (2011), who provide a comprehensive framework that includes chapters on risk assessment, hedging, mitigation, and management in global supply chains. An extensive and growing body of literature can be reached through this book. Haksöz, Seshadri, and Iyer (2011) devote a section to supply chain risk management on the Silk Road, covering interesting practice examples on supply chain risk from Asia, the Middle East, and Turkey. Lastly, Sodhi and Tang (2012) provide a good balance of academic and practitioner perspectives on supply chain risk as well as open problems in the field.

The initial incorporation of risk in supply chain management was through the supply chain contracts. To this end, Ritchken and Tapiero (1986) wrote a seminal paper that addresses the optimal design of options contracts that will meet the risk–reward preferences of a buyer in the presence of demand and price uncertainties. Li and Kouvelis (1999) study risk-sharing contracts with price uncertainty. To model risk aversion, supply chain management literature has generally used the preference-based utility maximization perspective proposed by Sandmo (1971). In this line of work, Eeckhoudt, Gollier, and Schlesinger (1995) first incorporate risk aversion in the standard news vendor model, and then various similar problems were studied by Agrawal and Seshadri (2000a, 2000b), Chen and Federgruen (2000), and Chen et al. (2006). The model introduced by McDonald and Siegel (1985), where a risk-adjusted valuation is developed for incomplete markets by using the market price of risk, was influential in supply chain literature. For example, Gaur and Seshadri (2005) are considered pioneers who introduced hedging with market instruments into the supply chain management literature. Moreover, recently, Gaur, Seshadri, and Subrahmanyam (2011) have addressed the value of postponement and early exercise to order and stock inventory. Demand and price risk that are correlated with the assets traded in the financial market are considered.

For commodities and commodity-type materials, managerial insights obtained in Wu, Kleindorfer, and Zhang (2002); Spinler, Huchzermeier, and Kleindorfer (2003); Seifert, Thonemann, and Hausman (2004); Martinez-de-Albéniz and Simchi-Levi (2005); Haksöz and Seshadri (2007); Haksöz and Kadam (2008, 2009); Haksöz and Şimşek (2010); and Sak and Haksöz (2011) are especially relevant. I refer interested readers to Haksöz and Seshadri (2007) for an overview of mathematical models for supply chain planning and management in the presence of spot and market exchanges. Most recently, in a book chapter by Haksöz and Seshadri (2011), risk management strategy of a manufacturer that can intelligently use the spot market as well as various financial instruments such as forwards, futures, and options is studied. Hedging joint price, basis, and yield risks are addressed in single- and multiperiod settings.

Scholars in supply chain management currently develop tools and methods to mitigate and hedge supply chain risks by operational hedging—real options (see, e.g., Haksöz and Seshadri, (2007), and references therein to get an overview of real options in supply chains) and flexibility tools—and financial hedging (e.g., forwards, futures, options, swaps).

Demand risks have been extensively studied, and risk-hedging tools have been developed. Inventory and capacity pooling, flexibility in resources and suppliers, and using multiple or backup suppliers are the key operational hedging strategies studied. See, for example, Eppen (1979); Eeckhoudt et al. (1995); Fisher and Raman (1996); Van Mieghem (1998, 2003); Van Mieghem and Rudi (2002); Netessine, Dobson, and Shumksy (2002); Corbett and Rajaram (2006); and Tang and Tomlin (2008) and references therein. Demand risk has been studied together with price uncertainty in the revenue management literature. For example, see Gallego and van Ryzin (1994) or Caldentey and Bitran (2003) and references therein to reach the vast literature on revenue management.

On the procurement and supply side of supply chains, price and market risks have been incorporated into the supply chain management models. Using spot markets and market exchanges to trade commodities and financial derivatives and integrating various real options into the supply contracts to mitigate various price, market, inventory, shortage, and supplier-related risks have been studied. To reach the key insights, see Li and Kouvelis (1999); Wu, Kleindorfer, and Zhang (2002); Kamrad and Siddique (2004); Goel and Gutierrez (2004); Gaur and Seshadri (2005); Martinez-de-Albéniz and Simchi-Levi (2005); Caldentey and Haugh (2006); Dong and Liu (2007); and Ding, Dong, and Kouvelis (2007) and references therein for a growing stream of research on operational and financial hedging of supply chain risks.

Finance literature on value-at-risk (VaR) aims to provide a single risk metric for financial loss over a given time period. Excellent reviews exist on this topic. See, for example, Tapiero (2004) for an excellent conceptual overview of key results. VaR is also used in operational decision making. For example, Tapiero (2005) addresses the inventory control problem ex post as a disappointment aversion problem, also known as a regret model. For example, see Bell (1985) and Gul (1991) and references therein for regret and disappointment models in decision making.

For empirical and managerial research on supply chain risk management, Hendricks and Singhal (2003) provide a great starting point. They examined the stock market reactions for various supply chain disruptions and showed that firms' stock prices are affected negatively with the news of supply chain disruptions. Chopra and Sodhi (2004) present an effective managerial perspective for supply chain disruption risk mitigation. Kleindorfer and Saad (2005) and references therein can be consulted for supply chain disruption management and robustness concepts. Tang (2006) reviews multifaceted modeling approaches on supply chain risk management.

Last but not least, I should note that risk hedging is an ex ante (before the fact) risk management method as opposed to recovering losses via insurance, which is ex post (after the fact) risk management (i.e., it occurs after the risky events have been realized). See especially Paté-Cornell (1996) and Tapiero (2004) for an integrated risk management approach that smoothly amalgamates ex ante and ex post risk management, which requires more research in the field of supply chain risk management.

Appendix B: Biography of Executives

Kordsa Global

Bülent Bozdoğan

Bülent Bozdoğan graduated from the Department of Management at Middle East Technical University in 1980. He started his career as an external auditor in PwC İstanbul (1980–1982). He joined Unilever Turkey in 1982 and worked for nine years as a manager in different financial and commercial positions. He then transferred to Brisa A.Ş, a joint venture of Sabancı Holding and Bridgestone Japan in 1991 and worked there until 2001. He was appointed chief financial officer of a global joint venture company (with DuPont) in the industrial yarn and tire reinforcement materials industry, located in Wilmington, Delaware. He was transferred to H.O. Sabancı Holding A.Ş in 2009 as head of Internal Audit and currently holds this position.

Arzu Öngün Ergene

Arzu Öngün Ergene was born in İstanbul in 1969. She graduated from Üsküdar American Academy for Girls in 1987 and then obtained a B.A. in marketing at Marmara University and an M.B.A. in international finance at Loyola University of Chicago. She started her career in the United States working as an accountant. Upon returning to İstanbul, she joined Eczacıbaşı Securities and then the Sabancı Group in 1994. She assumed several duties in different functions working as marketing and sales specialist, finance manager, and finance and accounting director. In 2009, she assumed the position of global sourcing director. In her current role, she is responsible for the sourcing activities of Kordsa Global for its eight plants around the globe.

Brisa

Fatih Tunçbilek

Fatih Tunçbilek is director of supply chain management at Brisa Bridgestone Sabancı Tyre Manufacturing and Trading Inc. He started his career as an industrial engineer at Brisa in 1988. After working for twenty-two years in the areas of plant operations, system development, and scheduling, in January 2010 he was put in charge of supply chain management. He also works with the supply chain risk management training programs for Enterprise Risk Management Association Turkey. He received a B.Sc. in industrial engineering from İstanbul Yıldız Technical University in 1986 and an M.Sc. in industrial engineering from Kocaeli University. His main focus areas are SCOR model implementation, supplier relation management, and supply chain risk management.

İdil Z. Taner Ertürk

İdil Z. Taner Ertürk obtained a B.S. in industrial engineering from İstanbul Technical and an M.S. in industrial engineering from Bosphorus University. She started working for Brisa Bridgestone Sabancı Tyre Manufacturing and Trading Inc. in 1999 in human resources. Later she was transferred to sales planning and has been working as a planning specialist and has remained there for eleven years. She is mainly responsible for master production planning and coordinating business continuity processes in the supply chain in Brisa.

AnadoluJet, Turkish Airlines

Sami Alan

Sami Alan graduated from Ankara Fen Lisesi (Ankara Science School) in 1993. He earned a bronze medal at the Math Olympic Games in Athens in 1992. He obtained a B.S. in industrial engineering from Boğaziçi University in 1999. He earned his M.B.A. program at the University of California Graduate School of Management in Irvine in 2001. He worked as senior analyst in revenue management at America West Airlines from 2001 to 2004. From 2001 to 2004, he served as president of the Turkish American Association of Arizona, the umbrella organization of the Turkish community in Arizona. He became advisor to the chief executive officer of Turkish Airlines in 2003 and was appointed vice president of revenue management in early 2004 and senior vice president of revenue management at the end of 2005. In 2008 he founded AnadoluJet, which now has the second-highest passenger count in the domestic market. He is a member of the SunExpress board of directors and has a private pilot license. He is also the president of the organization committee of the Archery Federation. As a serial entrepreneur, he is the founder or partner of a number of companies, including Japonbaligi.com, Suart, Onok, Okmeydani, IceNice, DuzcePark, AVMedya, and Turijobs.com.tr, whose areas vary from aquarium fish production to archery products sales and ice cream and from career websites and landscape plantation to advertising.

Notes

Prologue

1. This quote is taken from Mesnevi, the magnum opus of Mevlana Celaleddin Rumi, thirteenth-century Turkish Sufi. See for the Turkish version Can (2002a, 2002b); for selected pieces in English, a great selection by Barks et al. (1997) can be consulted. Translation by the author.
2. See, for example, http://www.goodreads.com/author/quotes/3137322.Fyodor_Dostoyevsky for a good selection of Dostoyevsky quotes. (Accessed November 29, 2012)
3. Refer to Chapter 2 for more details on risk intelligence.
4. This unique and original cover art as well as the other acrobat drawings in Part 2 were created by Professor Yankı Yazgan, a well-known Turkish psychiatrist and prolific scholar.

Chapter 1

1. Great quotes of Richard P. Feynman can be found, for example, at http://www.goodreads.com/.
2. See Muallimoğlu (1998 p. 228) for a great selection of Turkish proverbs and folk sayings.
3. See Google Books NGram Viewer at http://books.google.com/ngrams to run your own experiments in millions of books in different languages such as German, Russian, and Spanish. See also for the background research by Michel et al. (2010), called the "Culturonomics," more resources available at http://www.culturomics.org/Resources/A-users-guide-to-culturomics.
4. These Google searches were done on July 5, 2012. As you read this book, the numbers will of course have changed over time.
5. We also need to take into account the question if the term risk is used similarly in other languages such as in my native Turkish. This semantic difference would somewhat skew the distribution for "risk" vis-à-vis the implications the word possesses for other societies.

6. See the article at http://www.thebci.org/index.php?option=com_content&view=article&id=168&Itemid=256 (Accessed July 3, 2012).
7. The types of global disasters considered in this graph are as follows: drought; earthquake, epidemic; extreme temperature, flood; industrial accident, insect infestation, mass movement dry, mass movement wet, miscellaneous accident, storm, transport accident, volcano, and wildfire. The data is available at EM-DAT, the International Disaster Database.
8. Birol (2012). Also see the World Energy Outlook by the International Energy Agency to get insights on energy and climate change risks that impact the global supply chain networks. See http://www.worldenergyoutlook.org/ for details.
9. See WEF (2011) for a detailed study on the types of emerging risks and ways to manage them.
10. Global Risks Landscape 2012, World Economic Forum. Video available at http://www.weforum.org/videos/global-risks-2012-risk-landscape. See WEF (2012b).
11. See Casti (2010) for an illuminating analysis of social mood and its impact on world events of very different scales and scopes.
12. See Casti (2012) for an eye-opening tour de force of extreme events and how society can anticipate and increase its resilience.
13. Ibid, p. 299.
14. Toffler and Toffler (2006), p. 45.
15. See Bradenburger and Nalebuff (1996), who coined the term *co-opetition,* for a detailed strategic analysis of such relationships in global supply chain networks.
16. See Makridakis and Taleb (2009) and Makridakis, Hogarth, and Gaba (2009) for insightful treatments of increasing complexity and uncertainty in the business world as well as our inability to correctly forecast and control it.
17. I call these types of suppliers *keystone suppliers,* borrowing a term from ecosystem management. See Part 2 for curious similarities of global supply chain networks with biological ecosystems.
18. See Malik, Niemeyer, and Ruwadi (2011) for the perspective of McKinsey & Company on the future of supply chain management.
19. See Sheffi (2007) for an excellent overview of concepts and tools for supply chain vulnerability and resilience.
20. Ibid.
21. See Kleindorfer and Wind (2009) and Kunreuther (2009).
22. See Gomory (1995) for the details.
23. See Taleb (2007) for horizon-expanding work on the extreme and rare events. The black swan concept was popularized by Karl Popper in the context of falsifiability of assertions.
24. See Kambil and Mahidhar (2005).
25. See Brown, Chui, and Manyika (2011) for an overview of the big data and the challenges it creates for the future management of companies.
26. See Brynjolfsson, Hitt, and Kim (2011).
27. See Nagali et al. (2008) for details on how HP has developed a smart approach to mitigate and manage supply chain contract risks in its global network.
28. See, for example, the Zara case by Ferdows, Machuca, and Lewis (2002).
29. Numbers in the parentheses show the percentage of events in that respective sample.
30. "iPhone shortage: Supply woes or new model?" *The Montreal Gazette,* April 3, 2008, p. B7.

31. "January rain check for Wii Nintendo can't meet holiday demand," *The New York Daily News*, December 15, 2007, p. 45.
32. See the article at http://www.nytimes.com/2011/01/19/business/19boeing.html (Accessed July 3, 2012). Title: Boeing again Delays Delivery of Dreamliner.
33. See Chapter 8 and Chapter 9 in this book for risk intelligent supply chain examples in the global tire supply chain network, namely, Kordsa Global and Brisa.
34. "Goodyear strike tests firm's resolve: Must close plants to compete," *National Post's Financial Post and FP Investing*, Canada, October 13, 2006, p. Fp9.
35. "Harley strike in 2nd week," *The Washington Times*, February 10, 2007, p. C11.
36. See the news at http://www.nytimes.com/2007/08/15/business/worldbusiness/15imports.html?pagewanted=all. Title: Mattel Recalls 19 Million Toys Sent from China. (Accessed July 3, 2012) as well as my short opinion (Haksöz 2007).
37. See Sezer and Haksöz (2012) for a mathematical analysis of strategic timing of a product recall decision in a dynamic setting.
38. See Pinedo, Seshadri, Zemel (2002) for the famous Ford-Firestone Case Study that addresses the product recall risk problem in various dimensions such as quality, supplier relationships, product design, governmental reactions, and consumer perceptions.
39. See Haksöz (2008).
40. For details on operational risk modeling see, for example, Cruz (2002). See Hoffman (2002) for great practice examples of operational risk management. See Chernobai, Rachev, and Fabozzi (2007) for a good overview of operational risk in the context of Basel II. For an analysis of the breach of contract risk in the context of the supply chain contracts see, for example, Haksöz and Kadam (2009), Haksöz and Şimşek (2010), and Sak and Haksöz (2011). Also see Wagner, Bode, and Koziol (2009) for supplier default risk in the automotive industry.
41. Haksöz and Kadam (2008).
42. Snavely (2006).
43. See Haksöz and Seshadri (2007) for a comprehensive review of the literature that presents the links between spot market operations and supply chain management.
44. See Haksöz and Şimşek (2010) for the details of the bundled option design and its value in managing breach of contract risk. For general theory on financial derivatives and pricing, see Hull (2003).
45. Haksöz and Şimşek (2010).
46. See Mahoney (2005) for a good overview of breach of contract and contract remedies. As there are barriers to design efficient contracts, supply contracts need to be supported by damage measures. See Haksöz and Kadam (2009) for other details.
47. See the article in *The Economist,* available at http://www.economist.com/node/16846402 (Accessed July 3, 2012). Title: Making the Earth Move.
48. Haksöz and Şimşek (2010).
49. See Haksöz and Dağ (2012) for an application of weather derivative pricing in Turkey.
50. See Brockett, Wang, and Yang (2005) and references therein for a good overview of weather risk management.
51. See Barrieu and Scaillet (2010) for details. Also see Chen and Yano (2010) for a model of weather risk in supply chain planning.
52. See the article "Storms slow rail traffic, send coal loads down again" that appeared in the *Platts Coal Outlook,* Vol. 34 (8), February 22, 2010.
53. See the news report at http://www.aksam.com.tr/iste-iptal-edilen-ido-ve-ucak-seferleri-96050h.html (Accessed July 3, 2012). Title: İşre İptal Edilen İDO ve Uçak Seferleri.

54. See the website http://www.cityofchicago.org/city/en/depts/mayor/iframe/plow_tracker.html for this very beneficial application of real-time information sharing with citizens in Chicago, IL. (Accessed February 1, 2012).
55. See the original blog post by Haksöz (2012), available at http://cagrihaksoz.wordpress.com/.
56. See the original news report available at http://www.hurriyet.com.tr/ekonomi/18052579.asp (Accessed June 29, 2012). Title: Arilar Çalişmadi Kiraz Rekoltesi Yariya Düştü.
57. See the original news report at http://ekonomi.haberturk.com/makro-ekonomi/haber/712585-bu-yil-sevgililerin-yuzu-gulmeyecek (Accessed June 29, 2012). Title: Bu Yil Sevgililerin Yüzü Gülmeyecek.
58. See the details at http://en.wikipedia.org/wiki/2011_Thailand_floods (Accessed July 3, 2012).
59. See the full article at http://cporising.com/2012/02/13/supply-risk-will-continue-to-impact-companies-in-2012/ (Accessed July 3, 2012). Title: Supply Risk Will Continue to Impact Companies in 2012.
60. Ibid.
61. See the news article at http://www.guardian.co.uk/business/2012/feb/14/lloyds-thailand-flooding-2bn-dollars?CMP=twt_fd (Accessed July 3, 2012). Title: Thailand Flooding Costs Lloyd's of London US$2.2b.
62. Çakanyıldırım and Haksöz (2012).
63. See the example on the acronym BRIC-MIST http://www.guardian.co.uk/global-development/poverty-matters/2011/feb/01/emerging-economies-turkey-jim-oneill (Accessed July 1, 2012). Title: After BRIC Comes MIST, the acronym Turkey would certainly welcome.
64. See Çakanyıldırım and Haksöz (2012) for more details on the Turkish supply chains and logistics practices in Turkey.
65. See the article in *Forbes Magazine* at http://www.forbes.com/forbes/2010/0329/billionaires-2010-hong-kong-new-york-london-cost-of-living-large.html (Accessed July 3, 2012). Title: Cost of Living Large.
66. See the beautiful and inspiring book by Böhmer, Powell, and Atılhan (2008) for the dying tradition of nomadic life of Yörük Turks in Turkey.
67. According to historians, Türkmens (Oğuz Turks) have twenty-four tribes divided into two major classes: Bozok and Üçok. See Sümer (1999) for more details.
68. See Sümer (1999) and Findley (2005) for detailed historical accounts of Türkmen/Oğuz as well as the other Turkic tribes in Central Asia, the Middle East, and Anatolia as well as the wide steppes of today's Russia and Eastern Europe.
69. See Haksöz, Seshadri, and Iyer (2011) for details on how caravan trade is being practiced along the Silk Road supply chains.
70. See Faroqhi (1984) for other details on caravan trade and Ottoman towns in the sixteenth and seventeenth centuries. I am grateful to Kürşad U. Akpınar for introducing me to Faroqhi's work. This quote is Faroqhi (1984, p. 66).
71. Observing no risk event for some time in a process may create a false perception that safety is guaranteed in the future. Advocating this view, Feynman is quoted in Regester (2000).
72. Interested readers are referred to Haksöz (2012) for details.

73. This study is similar to recent risk perception studies in supply chains such as the one that compared the supply risk perceptions of procurement professionals in the United States and Germany. For example, see Zsidisin et al. (2008) and references therein.
74. For a recent overview of Toyota supply chain management, see Iyer, Seshadri, and Vasher (2009). For the general Toyota way of management, see Liker (2004).
75. See Isaka (1997) for a detailed report on how Toyota chose to use single sourcing for some of its suppliers but reconsidered this strategy after a fire crisis.
76. See Obrien (2002) for details.
77. See Narasimhan et al. (2009).
78. The lock-in phenomenon in economics goes back to Arthur (1994), who studied the dynamic lock-in process for a certain technology among competing technologies and its path-dependent consequences for the evolution of technology and innovations. See Chapter 2 in Arthur as well as Page (2006) in the context of path dependence.
79. See Perrett (2011) for the full story.
80. Dillon (2011).
81. See Mecham (2009) for the full story.
82. Ibid.
83. For more details, please see Haksöz (2012).
84. See Nishiguchi and Beaudet (1998) for the Aisin Seiki case.
85. For those who want to delve into scholarly papers on supply chain risk, key pointers to the literature are given in Appendix A.
86. See, for example, the insightful book by Anupindi et al. (2011) for the process view of operations and supply chain management. Such a view is one way of representing the types of risks observed in global supply chain networks. There may surely be other appropriate ways. The gist of the matter is to determine the risk drivers in the network and adequately manage them. See also Ritchie and Brindley (2007) for a useful framework to study supply chain risk management.
87. See Casti (2011) for a good overview of these four uncertainty types and the value of methods for management.
88. See Van Mieghem (2003) for a comprehensive review of operational hedging in the context of capacity management as well as Van Mieghem (1998) and Van Mieghem and Rudi (2002).
89. For a complete study on breach of contract risk and an intelligent method to mitigate it, see Haksöz and Şimşek (2010). For supply portfolio risk management, see Haksöz and Kadam (2009). Also see Dada, Petruzzi, and Schwarz (2007) for supplier reliability risk management.
90. For earlier work on supply chain contracts with options, see, for example, Barnes-Schuster, Bassok, and Anupindi (2002). Also see Haksöz and Seshadri (2007) for supply chain contract valuation with abandonment option. For an introductory study on supply chain contracts in general, see Cachon (2003).
91. See Behar (2008) on China and African mines. Also see Tevelson, Ross, and Paranikas (2007) for Boston Consulting Group's view on sourcing and risk hedging.
92. See Hagel and Brown (2005).
93. For more details, see Thomke (2003) and references therein.
94. For an application of weather derivative pricing in Turkey, see Haksöz and Dağ (2012).
95. For catastrophic risks and hedging instruments, see Chichilnisky (1996) and Michel-Kerjan and Morlaye (2007).
96. For an example on supplier auditing, see Kunreuther (2009).

97. Based on Michel et al. (2010).
98. The details of the price indices are as follows. The commodity food price index includes *cereal, vegetable oils, meat, seafood, bananas,* and *oranges price indices.* The commodity metals price index includes *copper, aluminum, iron ore, tin, nickel, zinc,* and *uranium price indices.* The commodity energy index includes the *crude oil, natural gas,* and *coal price indices.*

Chapter 2

1. Cook (1993), p. 421.
2. Ibid, p. 394.
3. See Bernstein (1996) for a tour-de-force work on the evolution of risk in history. Bernstein (1996, p. 110).
4. For a detailed report on what executives think about the rising complexity and the need for fresh perspectives, see WEF (2012a).
5. See De Bono (2004) for details on how concepts are an essential part of a beautiful mind. De Bono (2004, p. 121).
6. Michalko (2003).
7. See Gardner (1999) for more details.
8. See Buzan (2000) for details on these ten types of intelligence in practice.
9. See Hämäläinen and Saarinen (2007) for an interesting study on systems intelligence. Hämäläinen and Saarinen (2007, p. 3)
10. See De Bono (1985), p. 128.
11. See Ackoff (1999) for a selection of best management articles by legendary management scholar Russell Ackoff. This quote is Ackoff (1999, p. 171).
12. See, for example, http://www.goodreads.com/author/quotes/3137322.Fyodor_Dostoyevsky to read more Dostoyevsky quotes. (Accessed June 2, 2012.)
13. The Bosphorus is the famous strait that connects the Asian continent to the European continent.

Chapter 3

1. This quote is from Senge (1990) in his seminal work "The Fifth Discipline" on learning organizations. See Senge (1990, p. 67).
2. This quote is from Fung, Fung, and Wind (2008) for the beautifully told story of Li and Fung in the borderless globe. See Fung et al. (2008, p. 48).
3. From Kelly (1998), where networks are discussed from interesting angles in the new economy. See Kelly (1998, p. 46).
4. For example, see Crook (2009) for a good overview of complexity theory and networks. Also see Chapter 7, "The Origin of Wealth," by Beinhocker (2007).
5. See Erdös and Renyi (1959) for seminal work on random networks.
6. See, for example, Mandelbrot (1983) for details on fractals.
7. Ibid.

8. For the origins of the strength of weak ties argument, see Granovetter (1973). For an application in social networks and talent management, see Yakubovich and Burg (2009).
9. Yakubovich and Burg (2009).
10. See Page (2011), p. 25.
11. See, for example, Helbing (2010) for a succinct overview of anticipation and management of systemic risks in complex systems such as societies and economics.
12. Ibid.
13. See Holland (1995) for the pioneering work on emergence as well as Crook (2009) for a business perspective.
14. See McKelvey and Andriani (2010) for a nontechnical paper on complexity, emergence, and self-organized criticality.
15. See Bak (1996) for details on self-organized criticality and its connections to complexity on Earth. Bak (1996, p. 1).
16. Ibid.
17. Ibid.
18. Self-organized criticality is also known as self-induced criticality. See Helbing (2010) for an insightful discussion on the topic.
19. Helbing (2010).
20. Ibid., p. 7.
21. See Diamond (2005) for a tour-de-force on the dynamics of society collapse.
22. See Brunk (2002) for an interesting study on collapse of societies and its intricate connection to self-organized criticality.
23. See Bernstein (2007) for an interview with Gary Pisano of Harvard Business School on biotech supply chain networks.
24. Available at http://www.goodreads.com/author/quotes/12008.Peter_F_Drucker (Accessed August 12, 2012).
25. For a succinct and illuminating study on the value of networks, see Kelly (1998), p. 25.
26. See Arthur (1994) for a detailed study on increasing returns.
27. See Fung, Fung, and Wind (2008) for a tour-de-force of Li and Fung and its supply chain network orchestration.
28. Ibid.
29. Ibid., p. 37–38.
30. For the details on the impact of networks and customization, see Prahalad and Krishnan (2008).
31. From Kunreuther (2009), p. 396.
32. See Roe and Schulman (2008) for a great overview of high reliability management principles.
33. Ibid.
34. This loss figure is computed with an interoperability input–output model. See Crowther, Haimes, and Taub (2007) for more details on the analysis and the sectors affected directly and indirectly from Hurricanes Katrina as well as Rita.
35. These network interdependencies are caused by nonlinear couplings, as explained in Helbing (2010).
36. Ibid.
37. An example for such an approach can be seen in Sak and Haksöz (2011) for a supplier portfolio risk management where supplier defaults are dependent.

38. This discussion is based on the works by Kunreuther (2009) and Heal and Kunreuther (2005). Interested readers are referred to these papers and references therein.
40. To analyze such cases of terrorist versus target, game theoretic models are employed. See, for example, Heal and Kunreuther (2005).
41. For tipping point concept, see Schelling (1978) for a scholarly perspective and Gladwell (2000) for a popular one.
42. See Kunreuther (2009) as well as the Responsible Care Program of the American Chemistry Council at http://www.responsiblecare-us.com.
43. An earlier version of this section has appeared in my blog. Available at http://cagrihaksoz.wordpress.com/.
44. See the news article at http://online.wsj.com/article/SB10001424052970203920204577195121591806242.html (Accessed July 3, 2012). Title: New Risks for Nuclear Plants.
45. See Helbing (2010) for a list of measures to reduce network vulnerability.
46. See, for example, Tversky and Kahneman (1974) for details on perception biases.
47. See, for example, http://www.cbc.ca/news/world/story/2012/07/25/f-drought-usa-faq.html for the full details on U.S. drought (Accessed August 2, 2012). Title: Ten Things to Know about U.S. Drought.
48. See, for example, http://www.hurriyet.com.tr/ekonomi/21030475.asp for details. (Accessed August 2, 2012). Title: *By Kuraklik Baska.* Giftalyni kirip geairdi.
49. A number of production facilities of the companies covered in this book such as Kordsa Global and Brisa are located in this region of Turkey.
50. On June 20, 2012, a new map was published in Turkey showing the 326 active earthquake fault lines. For the details, see http://www.hurriyetdailynews.com/fault-lines-crisscrossing-turkey.aspx?pageID=238&nID=23606&NewsCatID=341 (Accessed June 22, 2012). Title: Fault Lines Crisscrossing Turkey.
51. See Haksöz, Seshadri, and Iyer (2011) for a diverse set of supply chain risk cases along the Silk Road, including the humanitarian logistics.
52. From Findley (2005) where a history of Turks ranging from economics, social life, and arts is thoroughly discussed. Findley (2005, p. 97).
53. For an overview of Turkish carpets and their features, especially the ones woven in different regions of Turkey such as Uşak under the Ottoman Turkish Empire, see Bloom and Blair (1997).
54. See Perrow (1999) for a pioneering book on how and why industrial accidents occur. This table is inspired by Perrow's work.

Chapter 4

1. Cook (1993), p. 443.
2. Ibid., p. 513.
3. Ibid., p. 499.
4. Lao Tzu is quoted in Hoeing (2000): "Failure is the foundation of success; success is the lurking place of failure." Thus, failure and success are the two sides of the same coin. Risk intelligent supply chains manage both success and failure with same diligence and vigor.

5. See Day and Schoemaker (2006) for an interesting treatise on weak signals and how they can be used strategically to gain competitive advantage in turbulent markets.
6. This quote is taken from De Bono (1985), p. 86.
7. The intricate dichotomy between individual and organizational learning is addressed by Peter Senge (1990).
8. Day and Schoemaker (2006).
9. See Massey and Wu (2005) for a discussion on underreaction and overreaction to information signals.
10. Ibid.
11. This classification is based on the general risk management framework suggested by Damodaran (2007).
12. For an interesting case of perception biases that affected an Everest Climb, see Roberto (2002).
13. There is a huge literature on perception and biases. Interested readers are referred to Slovic, Fischoff, and Lichtenstein (1982) for perception and biases in risk.
14. See Gray (2012) for details on managing natural disaster risks in global supply chain networks.
15. For details on these biases in the context of peripheral vision, see Day and Schoemaker (2006).
16. Slovic (1993, p. 225).
17. See Loewenstein et al. (2001) for an interesting study on the interplay of risk and feelings. Loewenstein et al. (2001, p. 270).
18. The full interview is in Madina (2008b). Also see details in Madina (2008b, p. 52).
19. See Sutton (2002) for a great discussion on the value of failures in organizational creativity. Sutton (2002, p. 17).
20. See Wind, Crook, and Gunther (2006) for strategic perspective on making sense of information. Wind et al. (2006, p. 99).
21. See Cannon and Edmondson (2005) for an insightful work on learning from intelligent failures.
22. Ibid.
23. Kelley and Littman (2001) provide the full story of IDEO.
24. Cannon and Edmondson (2005), p. 309.
25. Sutton (2002), p. 98–99.
26. See Edmondson (2011, p. 51) and also the April 2011 issue of *Harvard Business Review*, which focuses on failure from a diverse set of perspectives.
27. Some modeling approaches consider the supplier reliability risk in procurement decisions (see, e.g., Dada, Petruzzi, and Schwarz 2007; Haksöz and Kadam 2008, 2009; Tomlin 2009).
28. From Thomke (2003), p. 13.
29. See, for example, the Organisation for Economic Co-operation and Development (OECD) Workshop Discussion Document on "Chemical Accidents and Incidents" (Rosenthal et al., 2004). For a good introduction on near-miss management, see, for example, Mürmann and Oktem (2002), Phimister et al. (2003), and Oktem, Wong, and Oktem (2010).
30. See Phimister, Bier, and Kunreuther (2004) for a good overview of perspectives on accident precursor analysis and management.
31. See Corcoran (2004) for a good overview of precursors.

32. Kleindorfer et al. (2004) studied hazardous material accidents in chemical plants. They use a five-year historical accident database to come up with a methodology to identify the criticality of accidents with respect to their severity and frequency and thus connect these to the facility characteristics. Kirchsteiger (1997) using accident data collected in the European Union from 1984–1995 shows that using precursor information to estimate the accident probabilities is valuable using accident data collected in the European Union from 1984 to 1995. The method he proposes uses the Bayesian updating procedure for conjugate distributions for the precursor frequency and the conditional density of a major accident given a precursor. Elliot et al. (2008) analyze accident data collected from 1996 to 2000 by the U.S. Environmental Protection Agency as well as the occupational illness and injuries reported by the Occupational Safety and Health Administration. No significant support is shown that low occupational illness and injuries are sufficient for low-probability/high-severity risks.
33. See Cruz (2002), Hoffman (2002), and Chernobai, Rachev, and Fabozzi (2007) for details and interesting applications of operational risk management; see Cruz and Pinedo (2008) for a tutorial on quality management and near-miss reporting in the realm of operational risk in service industries such as finance, health care, transportation, and hospitality.
34. See March, Sproull, and Tamuz (1991) and Carmeli and Schaubroeck (2008) for interesting studies on learning from failures when the sample size is very small, even one.
35. Lampel and Shapira (2001) study strategic surprises and show that using early warning systems in managerial and organizational management may mitigate strategic surprises.
36. See Bier et al. (1999) for a good survey of extreme risk assessment and management approaches.
37. See Loewenstein et al. (2001) for more details on the effect of feelings on risk taking.
38. See Casti (2011), p. 2.
39. Ibid., p. 3.
40. For full details, see ibid., p. 4–5.
41. Ibid., p. 4.
42. In the context of operational and credit risk management, power law distributions are addressed in, for example, Cruz (2002), Panjer (2006), and McNeil, Frey, and Embrechts (2005).
43. See Andriani and McKelvey (2009) for a good overview of scale-free theories and Paretian thinking models in organization research.
44. See, for example, Mandelbrot and Hudson (2005) for an excellent overview of fractal thinking and power laws in financial markets.
45. Andriani and McKelvey (2009).
46. Andriani and McKelvey (2009).
47. Ibid.
48. Ibid.
49. Ibid.
50. Casti (2011).
51. See Fung, Fung, and Wind (2008) for a successful application at Li and Fung.
52. See La Porte (1996) for a good overview of high reliability organizations. La Porte (1996, p. 61).
53. Available at http://www.goodreads.com/author/quotes/4012.Wilkie_Collins (Accessed August 12, 2012).

54. See www.goodreads.com for great selection of quotes.
55. To this end, Thomas Jefferson is quoted in Son and Kornell (2010) as saying, "He who knows best knows how little he knows." People's knowledge and uncertainty about their own knowledge is addressed in ibid.
56. See, for example, Roberto (2002) for a great story on mountaineers who were set to climb Mt. Everest. Once various biases interfere in the process, decision-making and risk-taking processes get debilitated.
57. See Kunreuther and Slovic (1999) for an interesting study on stigmatization of risks and the vulnerability it creates for society.
58. See McKelvey and Andriani (2010) for a detailed discussion on these organizations and how it relates to resilience.
59. Charles Doswell is a professional weather forecaster as quoted in Nadin (2006), p. 30.
60. See Paté-Cornell (1986).
61. Bier and Mosleh (1990) look at the accident precursors and how the precursor frequency changes the accident frequency in a Bayesian updating framework. One can also analyze a system where there are multiple precursors. Interdependence between the precursors can be analyzed by copula methods; see Yi and Bier (1998).
62. Thomke (2003), p. 166.
63. Surely, Six Sigma methods can also be used to monitor deviations from the norm in supply chain processes. Six Sigma methodology is also shown to be valuable in operational control in global asset management. See Pinedo (2010) for a recent book presenting a diverse set of academic and practitioner views.
64. Being mindful requires catching early warning signals of troubles. See Langer (1989) for a complete discussion on mindfulness. Langer (1989, p. 134).
65. I am grateful to my dear father, Mehmet Haksöz, who is the original creator of this concept.
66. These critical transitions correspond to bifurcations. For a good review, see Scheffer et al. (2009) and Scheffer (2009). For earlier work, see Scheffer et al. (2001).
67. See Scheffer (2009) for more detailed examples of critical transitions in lakes, climate, oceans, terrestrial ecosystems, and human societies.
68. See Scheffer (2009) for details on critical transitions and early warning signals.
69. Scheffer et al. (2009).
70. Ibid.
71. Roughly speaking, a basin of attraction is defined as the set of initial conditions leading to long-term behavior that approaches a particular attractor. An attractor is a subset of the state space in a dynamical system to which there is attraction in time. See, for example, http://www.scholarpedia.org/article/Basin_of_attraction for details on attractors and basins of attraction in dynamical systems.
72. Scheffer (2009).
73. See Sornette (2006), who wrote an excellent detailed study of critical phenomena. For critical crashes, see Johansen, Sornette, and Ledoit (1999), Sonette (2002, 2003), and Osorio et al. (2009) for intimate connections between epileptic seizures and earthquakes.
74. See, for example, methods proposed in Bier and Mosleh (1990) and Paté-Cornell (2002).

75. False positive and false negative are Type I and II errors, respectively, in statistics. For more on this topic and how they are used in detecting the precursors effectively, see, for example, Paté-Cornell (2004).
76. See Paté-Cornell (1996).
77. See Damodaran (2007) for details on VaR (Value at Risk) methods.
78. See Haksöz and Kadam (2009) and Sak and Haksöz (2011) for the mathematical details of this risk measure Supply-at-Risk (SaR) as well as Conditional Supply-at-Risk (CSaR).
79. Although black swans are a rare sighting, black swans have been seen (although rarely) among white ones in various artificial and natural lakes in İstanbul.
80. This photo was taken on the pretty lake at Sabancı University in İstanbul, Turkey. In this photo, there are four different species. The careful eye will observe three barnacle geese, which normally do not live in Turkey apart from the migration periods. However, in summer 2012, this small group in the photo decided to call Tuzla, İstanbul—instead of northern Europe—its summer home.

Chapter 5

1. See the transcript of the full conversation with Joi Ito at http://www.edge.org/conversation/innovation-on-the-edges (Accessed June 30, 2012).
2. Cook (1993), p. 537.
3. Available at http://en.wikiquote.org/wiki/Isaac_Asimov (Accessed August 13, 2012).
4. Available at http://www.goodreads.com/author/quotes/12793.Charles_Darwin (Accessed August 12, 2012).
5. When we examine the word *improvisation*, its root goes back to Latin *improvisus*, which means "not seen ahead of time."
6. For an eye-opening article on improvisation and creativity in jazz and organizations, see Barrett (1998), p. 606.
7. See Perry (1991) for the improvisational approach to strategy, p. 51.
8. See Weick (1998) for a good overview of improvisation in organizations, p. 544.
9. Ibid. Also see Berliner (1994) for intricate details on jazz and improvisation.
10. See Manisaligil and Haksöz (2012) for a discussion on risk and improvisation in Turkish classical and folk music genres.
11. See Ross (2011) for details.
12. For an insightful discussion on resilience and resilient people, see Coutu (2002, p. 47).
13. Scheffer (2009).
14. See Weick, Sutcliffe, and Obstfeld (1999).
15. Coutu (2002).
16. Ibid., p. 52.
17. Ibid., p. 48.
18. A great book by Shekerjian (1990) studies the path to genius from the eyes of the forty MacArthur fellows, p. 202.
19. See Weick and Sutcliffe (2007, p. 71).
20. See Sheffi (2007) for one of the earliest works on resilience in supply chains.
21. See Pettit, Fiksel, and Croxton (2010) for a conceptual supply chain resilience model.

22. See Carpenter et al. (2001) for more details on the adaptive cycle framework. For a layman's treatment of the adaptive cycle, see Walker and Salt (2006).
23. See also Linnenluecke and Griffiths (2010) for an application of resilience thinking on climate change.
24. See Walker and Salt (2006) for more details on resilience thinking.
25. Ibid.
26. Ibid.
27. Ibid.
28. From Ohmae (1982), who had a great impact on strategic thinking, p. 276.
29. See Ridderstrale and Nordstrom (2007), who provide a provocative approach to today's battle of brains.
30. See Florida (2002) for a tour-de-force of the creative class and its provocative implications for the globe.
31. See BrainyQuote for other interesting quotes of Steve Jobs. Available at http://www.brainyquote.com/quotes/quotes/s/stevejobs416925.html (Accessed June 17, 2012).
32. See, for example, Thomke (2010).
33. Ibid.
34. See De Bono (1991) for more details on these thinking styles, p. 23.
35. Ibid.
36. For a detailed discussion on the defocused attention and other characteristics of creative individuals, see Simonton (1999), p. 90.
37. See De Bono (1985), who shows how different people (from business executives to mountaineers) define and discuss their success, p. 177.
38. See Kelley and Littman (2001) for an insightful story of IDEO.
39. This quote is taken from Shekerjian (1990), p. 66.
40. These definitions are given in http://dictionary.reference.com (Accessed July 29, 2012).
41. See Jakob (1977) for a great article on tinkering in the context of biological evolution.
42. Ibid., p. 1163.
43. For a detailed discussion on adjacent possible, see Kauffman (2000) and Johnson (2010).
44. Kauffman (2000), p. 229.
45. See Johnson (2010) for a detailed discussion on adjacent possible and how it explains the process of creating new ideas, p. 42.
46. Jakob (1977).
47. See Dörner (1996) for an interesting study on identifying and managing failures in complex environments, p. 45.
48. See Weick and Sutcliffe (2007) for more details on mindfulness and its significance in managing the unexpected, p. 32.
49. Ibid.
50. See Langer (1989) for a tour-de-force treatment of mindfulness as well as mindlessness, p. 62.
51. See the original article by Lindblom (1959) on muddling through for the details.
52. See Simon (1962).
53. See Kay (2011) for an insightful book on oblique problem solving, p. 57.
54. Available at www.brainyquote.com/quotes/authors/m/malcolm_forbes.html (Accessed June 2, 2012.)
55. Available at www.brainyquote.com/quotes/authors/m/maya_angelou.html (Accessed June 2, 2012.)

56. See http://oxforddictionaries.com/definition/english/diversity?q=diversity (Accessed July 10, 2012).
57. See Hayes et al. (2005) for details on corporate operations strategy and how diversification fits into it.
58. See Page (2011) for an illuminating book on diversity and complexity.
59. See Norberg and Cumming (2008) for a great book on a diverse set of perspectives on complexity theory.
60. See Scheffer (2009) for details on species and types of biodiversity.
61. See Levin (1999) for information on the biodiversity, fragility, and resilience of ecosystems.
62. See Scheffer et al. (2001) for a great discussion on catastrophic shifts in ecosystems. For an overview of critical transitions and regime shifts that are catastrophic, Scheffer (2009) is highly recommended.
63. Levin (1999), p. 167.
64. Scheffer (2009).
65. See May, Levin, and Sugihara (2008).
66. See Levin (1999) for details on modularization and its impact on ecosystem resilience.
67. See Page (2011) on how modularity helps increase system robustness.
68. See Page (2007) for another great book on how diversity is powerful in creating better groups, firms, schools, and societies.
69. See Surowiecki (2004).
70. See Popova (2012). The 1962 video showing the elevator experiment can be seen on the same website.
71. See Page (2007), p. 209.
72. See Dye (2008) for a practical approach to prediction markets.
73. See http://www.hsx.com/ for details.
74. See Page (2007).
75. See *Economist* (2012).
76. Ibid.
77. See Carr (2007) for details on crowd ignorance in the context of open-source innovation, p. 4–5.
78. See Ashby (1956) for the original work as well as Page (2011) for a succinct discussion.
79. See Page (2011).
80. See Weick and Sutcliffe (2007) for a discussion on how the law of requisite variety helps explain the high reliability of organizations' management of complex environmental disturbances.
81. See Casti (2012).
82. See Bar-Yam (2004) for details on the multiscale law of requisite variety.
83. Ibid.
84. See Page (2011) for a detailed discussion on redundancy and diversity.
85. Levin (1999), p. 202–203.
86. Ibid.
87. Ibid.
88. See McCann (2000) for a great discussion on the diversity–stability debate, p. 228.
89. See Kauffman (2000) for a tour-de-force of autonomous agents and their co-creation.
90. Ibid., p. 238.
91. Lao Tzu's quotes are available at http://www.goodreads.com/quotes/ (Accessed August 12, 2012).

92. Mathematically, all of these flexibility strategies to mitigate various supply chain risks are shown to behave in a concave manner.
93. See Tang and Tomlin (2008) for details.
94. See Lee (2004) for the details of the relationship between Hewlett-Packard and Canon.
95. See also Jordan and Graves (1995) and Graves and Tomlin (2003) for in-depth studies of process flexibility in single- and multiple-stage settings for different manufacturing configurations.
96. See Fung, Fung, and Wind (2008).
97. See Lamarre, Pergler, and Vainberg (2009) for McKinsey & Company's perspectvie on network flexibility.
98. For various models and applications of real options in global supply chain networks, see, for example, Kogut and Kulatilaka (1994), Li and Kouvelis (1999), Kouvelis, Axarloglou, and Sinha (2001), Ding, Dong, and Kouvelis (2007), and references therein.
99. See Vanderhaeghe and de Treville (2003) for details on how flexibility can be practiced as well as malpracticed at different levels of a supply chain network.
100. A supplier portfolio can also include short-, medium-, and long-term suppliers in local, regional, and global locations.
101. See Kauffman (2000) for details on diversity and phase transitions.

Chapter 6

1. See Dörner (1999).
2. See www.quoted6.com/quotes/440. Accessed November 28, 2012.)
3. See www.quotationsbook.com/quote/34701. (Accessed November 28, 2012.)
4. See Rilke (1967), p. 234.
5. See Albeverio, Jentsch, and Kantz (2010) for interesting perspectives on extreme events.
6. This is similar to having both upsides and downsides in Chinese. As used frequently especially by politicians, risk in Chinese is characterized by two concepts: danger (downside) and opportunity (upside).
7. See Sornette (2006a) for more details.
8. Sornette (2006b) studies endogenous versus exogenous shocks in social networks.
9. See Casti (2010) for an interesting book on the social mood and its peculiar global effects on different time scales.
10. See Sornette (2002a) for details on innovation with blockbusters in the movie and pharmaceutical industries.
11. See De Vany and Walls (1996) for details on motion picture dynamics. Information cascade is very similar to risk cascading, where feedbacks among different events create a higher risk for the system, discussed in the Integrator role.
12. See Waldrop (1992) for a tour-de-force on complexity science for lay readers, p. 126.
13. Refer to Chapter 5 for more discussion on mindfulness.
14. See Rock (2009) for a good overview of social brain in the context of management.
15. See Michalko (2003).
16. Ibid.

17. See for more details http://the99percent.com/tips/7146/Tripping-into-Terra-Incognita-How-Mistakes-Take-Us-To-New-Places (Accessed March 8, 2012).
18. See Simon (1986) for an intriguing contribution for the managerial creativity process and its apparent connection to problem solving.
19. See Rivkin (2000) for an insightful analysis of complex business strategies under the threat of imitation, p. 833–834.
20. See Arthur (2009) for a great story on how technology evolves.
21. For an interesting discussion of growth in civilizations and the value of diversity, see Page (2007), p. 317.
22. See Wind, Crook, and Gunther (2006) for details on the practice of flying upside down.
23. Ibid.
24. See Shekerjian (1990), p. 91.
25. See Rijpma (1997) for details on complexity, learning in the context of normal accidents, and high reliability theories.
26. Ibid.
27. See Sornette (2008) for details on bubbles and breakthroughs, p. 176.
28. Ibid.
29. Apple announced the list of its suppliers in 2011; it shows great diversity. Without the technological improvements and the capacity investment of these suppliers for the past few decades, a blockbuster success would not be probable. See http://www.tomsguide.com/us/Apple-Supplier-Responsibility-Suppliers-2011-Foxconn-Underage-Labor,news-13895.html for details (Accessed August 4, 2012).
30. See Zander and Zander (2000), p. 101.
31. See Pascale, Sternin, and Sternin (2010) for illuminating stories of positive deviance.
32. Ibid., p. 13.
33. See the website for the Positive Deviance Initiative for success stories of the PD approach (http://www.positivedeviance.org).
34. Pascale et al. (2010).
35. For an interesting story of a surgeon in unusual, risky, and complex environments, see Gawande (2007), p. 251–257.
36. In this process, appreciative inquiry methods can be effectively employed. See, for example, Cooperrider and Srivastva (1987).
37. I want to acknowledge my colleagues Murat Çokol and Aydın Albayrak for introducing me to the peculiar yet admirable lives of extremophiles.
38. From http://www.brainyquote.com/quotes/quotes/f/friedrichn101616.html (Accessed August 5, 2012).
39. See Persidis (1998) for a good overview of extremophiles.
40. See Appleyard (2012) for a recent story on extremophiles that appeared in the mass media. Alternatively, see www.moreintelligentlife.com/content/ideas/some-it-very-hot. (Accessed November 28, 2012.)
41. For a list of applications, see Persidis (1998).
42. See Clegg (2002) for a good overview of business lessons from extremophiles.
43. Bradenburger and Nalebuff (1997).
44. Clegg (2002).
45. See Kay (2011) for a discussion on monoculture versus plural cultures and the implications for adaptation and flexibility.
46. See Sornette (2009) for a detailed discussion on dragon-kings. For an application in epileptic seizures, see Osorio et al. (2009).

47. Sornette (2009).
48. Oscillations exist in transiently disabling seizures in epilepsy as well as sudden, aperiodic earthquakes. Both systems have great similarities even though in completely different contexts. See Osorio et al. (2009) for details.
49. Osorio et al. (2009).
50. See Levinthal (1997) for a classical reading on adaptation and strategy.
51. Rivkin (2000).
52. For details on the Mt. Everest expedition and how the system complexity played a critical role, see Roberto (2002).
53. Taken from a great selection of quotes in Cook (1993), p. 439.
54. For a great book on strategic risk taking, see Damodaran (2007).
55. Simon (1986), p. 17.
56. See Schwalbe (2012).
57. For an insightful work on technological progress and innovation in different societies, see Mokyr (1990), p. 158.
58. See Shekerjian (1990) for more on risk-taking behaviors of MacArthur fellows, p. 24.
59. See, for example, Machowiak (2008) for a practical method.
60. Loewenstein et al. (2001).
61. I acknowledge my colleague Wojciech Machowiak of Poznan School of Logistics for sharing this interesting proverb.
62. Dörner (1996).
63. See also Sornette (2008) for the limitations of control we have over success.
64. See Dörner (1996) for an interesting discussion on ballistic versus nonballistic behavior in complex systems, p. 179.
65. Mokyr (1990), p. 158.
66. See also Ben-Haim (2006), who aims to optimize robustness to failure under severe uncertainty.
67. See en.wikiquote.org/wiki/Abraham_Lincoln. (Accessed November 28, 2012.)
68. For effective risk communication, see, for example, Stone, Yates, and Parker (1994).
69. See http://nationalzoo.si.edu/Animals/AfricanSavanna/fact-cheetah.cfm for other factual details on this amazing cat (Accessed August 7, 2012).
70. See *The Encyclopedia of Animals: A Complete Visual Guide* for interesting facts of the animal kingdom that shares this planet with us.
71. See Florida (2002) for a great discussion on creative and noncreative cities.
72. See Sornette (2009) and Osorio et al. (2009) for more details on this qualitative phase diagram.

Chapter 7

1. See Tripp (1970) for a great selection of quotes, p. 694.
2. Ibid., p. 697.
3. See Can (2002b), p. 599
4. Ibid., p. 695.
5. The *Oxford English Dictionary* is available online at http://oxforddictionaries.com (Accessed June 11, 2012).
6. See De Bono (1991), p. 167.
7. See www.quoteworld.org/quotes/8232. (Accessed November 28, 2012.)

Chapter 8

1. Mr. Sakıp Sabancı, a legendary businessman in Turkey who moved the Sabancı Group to the global stage. See http://en.wikipedia.org/wiki/Sak%C4%B1p_Sabanc%C4%B1 for details of his biography.
2. See Kordsa Global 2010 Annual Report for details. Available at http://www.kordsaglobal.com/media/downloads/faaliyet_raporlari/2010_annual_report.pdf (Accessed April 30, 2012).
3. See "Tire Cord—Global Strategic Business Report 2010" by *M2presswireNewspaper Source Plus.*
4. Kordsa Global 2010 Annual Report. http://www.kordsaglobal.com/media/downloads/faaliyet_raporlari/2010_annual_report.pdf. (Accessed April 30, 2012.)
5. See *Kordsa Global Endustriyel Iplik ve Kord Bezi Sanayi ve Ticaret A.S. 2011*, SeeNews Research & Profiles (Company Profiles), August 1–8.
6. See Public Disclosure Platform, *Kordsa Global Endüstriyel İplik ve Kord Bezi Sanayi ve Ticaret A.Ş.*, http://www.kap.gov.tr/yay/English/Sirket/sirket.aspx?sirketId=1009 (Accessed April 30, 2012).
7. Thomson Reuters, "Kordsa Global Endustriyel Iplik ve Kord Bezi Sanayi ve Ticaret AS." http://www.reuters.com/finance/stocks/overview?symbol=KORDS.IS (Accessed April 30, 2012).
8. *Kordsa Global Endustriyel Iplik ve Kord Bezi Sanayi ve Ticaret A.S. 2011*, See News Research & Profiles (Company Profiles), August 1–8.
9. Ibid.
10. Kordsa Global 2010 Annual Report. http://www.kordsaglobal.com/media/downloads/faaliyet_raporlari/2010_annual_report.pdf Also, see a corporate video available at http://www.youtube.com/watch?v=zTapEoteOWo (Accessed June 28, 2012).
11. Kordsa and Kordsa Global are used interchangeably throughout this book.
12. See the global supply chain network map for Kordsa Global in Figure 8.1.
13. At the time of this conversation, in April 2011, the crude oil price was hovering around US$125/barrel. As of May 2012, it went down to US$110/barrel.
14. See Pinedo, Seshadri, and Zemel (2001) for the famous Ford-Firestone case study.
15. The spot price for cotton hit 216.62 U.S. cents per pound in April 2011 when this interview was conducted. As of May 22, 2012, the spot price dropped to 70.72 U.S. cents per pound. Global cotton production has increased in the last year due to high prices, which eventually created a glut. This volatility needs to be anticipated and managed in this market. (Source: http://www.indexmundi.com/commodities/?commodity=cotton&months=60)
16. See Chapter 9 for a detailed discussion of Brisa and its risk intelligent supply chain.
17. Also see the detailed discussion on this topic in Chapter 9.
18. See Crowther, Haimes, and Taub (2007) for a study that addresses the direct and indirect economic impact of Hurricane Katrina on various sectors such as oil and gas. The cascading losses have been estimated at US$800 million to the State of Louisiana for the first month following the disaster.
19. See also the discussion on earthquake risk management of Brisa in Chapter 9.
20. See Haksöz and Şimşek (2010) for an interesting study that models the renegotiation process as a real option product (i.e., a price renegotiation option) and shows its value in mitigating the breach of contract risk in supply chain contracts.

21. Simultaneous zooming in and out is discussed in the book by Wind, Crook, and Gunther (2006). With this technique, one can see where one is going and the path to get there at the same time.
22. See, for example, Grove (1996).
23. See also the discussion on natural rubber in Chapter 9 within the context of Brisa.
24. See Tang and Tomlin (2008) for the value of supplier portfolio flexibility and the discussion in Chapter 5.

Chapter 9

1. For complete details, see the industry profile report "Global Tires and Rubber," *Datamonitor*, June 2011.
2. Ibid.
3. Global market share data are obtained from the Bridgestone Europe website, http://www.bridgestone.eu/corporate/about-bridgestone/rubber-and-tyre-market/world-tyre-market-share (Accessed June 16, 2012).
4. See "Bridgestone Data 2011," http://www.bridgestone.co.jp/corporate/library/pdf/BSDATA2011.pdf (Accessed April 22, 2012).
5. See "Bridgestone Corporation Annual Report 2011," http://www.bridgestone.com/corporate/library/annual_report/pdf/bs_annual_2011_operational.pdf (Accessed April 22, 2012).
6. See "Brisa Milestones" for all the details of awards, honors, and other interesting events in a timeline, http://www.brisa.com.tr/English/Brisa/Corporate/Milestones.aspx (Accessed June 16, 2012).
7. See the report "Brisa," İş Investment, http://www.isyatirim.com.tr/WebMailer/files_att/2_20110208103930395_1.pdf (Accessed April 22, 2012).
8. Ibid.
9. "*Brisa Bridgestone Sabancı Lastik Sanayi ve Ticaret A.Ş*," See News Research & Profiles (Company Profiles), August 1–5.
10. "Brisa."
11. *Brisa Bridgestone Sabancı Lastik Sanayi ve Ticaret A.Ş*.
12. Ibid.
13. The original Turkish phrase was used: *Yola Güvenle Çık, Yolun Hep Açık*.
14. Ibid.
15. See http://www.ise.org/ for details.
16. See http://www.bridgestone.com.tr for details on Brisa's global brand and how it is being managed in the global supply chain network.
17. See http://www.lassa.com.tr for more details on this specific tire brand from Brisa.
18. See www.otopratik.com.tr for the service brand of Brisa and the different services provided.
19. For the details of the Retreading Process at Bandag, please see http://www.bandag.eu/bandag-eu/English/A-Bandag-is-always-a-Bandag/The-Bandag-Precure-Retreading-Process.html (Accessed June 4, 2012).
20. For details of the SCOR Model, see http://supply-chain.org/scor (Accessed June 4, 2012).
21. See Chapter 8 for details on Kordsa Global's risk intelligent supply chain.

22. See Haksöz and Seshadri (2007) for details of this model.
23. See Haksöz and Seshadri (2007) and Haksöz and Şimşek (2010) for detailed analyses of such flexible supply chain contracts.
24. Bowden (2010) gives a detailed report on maritime piracy risks and the economic costs along global supply chain networks.
25. Refer to Table 9.1 for a complete list of supply chain risks and related strategy and course of action to manage those risks.
26. On June 20, 2012, a new map was published in Turkey showing the 326 active earthquake fault lines. Such a detailed map could help Brisa and other firms calculate their earthquake risk exposures much more accurately. See http://www.hurriyetdailynews.com/fault-lines-crisscrossing-turkey.aspx?pageID=238&nID=23606&NewsCatID=341 (Accessed June 22, 2012).
27. Also refer to Chapter 8 on Kordsa Global to see how contracts are being managed on the upstream supply chain.
28. See, for example, Haksöz and Şimşek (2010) and Sak and Haksöz (2011) for more details on breach of contract risk and effective methods to assess and mitigate this risk.
29. For details of how Apple and Dell took actions in the 1999 Taiwan earthquake, see Sheffi (2007).
30. For more details on CDP and CDP Turkey-100 report, visit https://www.cdproject.net/en-US/WhatWeDo/Pages/Turkey.aspx (Accessed June 12, 2012).
31. For more information on GDI, visit https://www.globalreporting.org (Accessed June 12, 2012).
32. For example, the catastrophic floods in Thailand during July 2011 to January 2012 hit a record of US$45.7 billion in damages according to the World Bank. See http://en.wikipedia.org/wiki/2011_Thailand_floods for a complete report on the Thailand flood (Accessed June 14, 2012).
33. Overstock and understock risks occur when the market demand does not match the inventory on hand. That is, you may have more inventory (overstock) than you need or less inventory (understock) than required.
34. Brainstorming using imagination on potential threats could reveal *black swans* for a company and supply chain network.
35. See, for example, Weick and Sutcliffe (2007) for interesting examples of how postmortem analysis can be conducted and how this analysis can improve the mindfulness of organizations.
36. These risk initiators play the role of *supply chain network healers* discussed in Chapter 6 in the context of extremophiles.

Chapter 10

1. THY is the acronym used for *Türk Hava Yolları*, or Turkish Airlines.
2. The city of Antalya is one of the most popular tourist destinations in Turkey on the Mediterranean coast.
3. İstanbul was also one of the major trading and supply chain hubs on the historical Silk Road for centuries. See Haksöz, Seshadri, and Iyer (2011), who propose a novel approach to relate the supply chain strategies and practices of the historical and modern Silk Road.

4. It has been reported that on July 7, 2012, a historic record was set for İstanbul Atatürk Airport, with 1,119 flights and 124,380 passengers in 24 hours. See http://www.dha.com.tr/ataturk-havalimaninda-rekor_336478.html (Accessed July 17, 2012).
5. In ancient history, Anatolia and the Asia Minor were used to denote the Asian part of Turkish Republic. In Turkish, this part of Turkey is named Anadolu as also used similarly under the Seljuki Turkish and Ottoman Turkish Empires. See also Haksöz et al. (2011) for the historic Silk Road in Anatolia and the management of supply chains on the Silk Road.
6. See Hammer (2004) for various types of operational innovations that make or break companies.
7. Gaziantep is located in southeastern Turkey, whereas Trabzon is in the eastern Black Sea region. Normally, this trip would take more than ten hours by bus.
8. See, for example, Lederer and Nambimadom (1998) for a technical analysis of airline networks.
9. Zara manages its global supply chain network using a fixed bus-like distribution schedule that enables highly synchronized operations. See Ferdows et al. (2002) for more details.
10. The Turkish phrase used for this motto was "*Tam Kadro Sabiha Gökçen'deyiz.*"
11. Adding one strategic route to the portfolio of flights enhances the value of the flight portfolio. This strategy has been found successful in other industries such as automotive. Provision of limited strategic flexibility incurring marginal cost goes a long way in satisfying most of the customers. Interested readers are referred to Jordan and Graves (1995). Also see the section on flexibility in Chapter 5.
12. Satisfying customers at the fringes or tails has been a topic of great interest. Anderson (2008) is the main promoter of such a concept. This is achieved by playing the Ingenious role in the I-Quartet model and betting on positive black swans.
13. This strategic choice reminds one of the effective management of Singapore Airlines. See Heracleous and Wirtz (2010) for a perspective on this airline's innovative strategy.
14. For more details on the peripheral vision in the context of risk intelligent supply chains, refer to Chapter 4. Also see Day and Schoemaker (2006) for a general discussion.
15. Alan likens the selection of talent based on competence and capability to the Ottoman Turkish Enderun school system, where gifted Christian youngsters were trained and educated at the palace for higher academic, bureaucratic, and military positions. See http://en.wikipedia.org/wiki/Enderun_School (Accessed June 16, 2012).
16. See Howe and Strauss (2007) for a discussion of this generation and implications for talent management.
17. See Buyck (2009).
18. Private communication with Sami Alan, Fall 2011.

Epilogue

1. From Cook (1993), p. 30.
2. Ibid., p. 208.
3. Ibid., p. 421.

4. These seeds are called the drift seeds, which ride the ocean currents and travel long distances. For example, the best-known plant drifter is the coconut seen on tropical beaches. See http://waynesword.palomar.edu/pldec398.htm for a captivating discussion on these types of seeds and their unique lives. (Accessed August 7, 2012.)
5. See Giegengack and Bordeaux (2009) for more details on honeybees and ants and how they manage intricate information networks with ease and efficiency.

Appendix A

1. More details on this research network can be found online at http://www.tlog.lth.se/forskning/iscrim.

Bibliography

Ackoff, R. L. 1999. *Ackoff's Best: His Classic Writings on Management.* New York: John Wiley & Sons.

Agrawal, V. and S. Seshadri. 2000a. Effect of risk aversion on pricing and order quantity decisions. *Manufacturing and Service Operations Management* 2(4):410–423.

Agrawal, V. and S. Seshadri. 2000b. Risk intermediation in supply chains. *IIE Transactions* 32(9):819–831.

Albeverio, S., V. Jentsch, and H. Kantz (Eds.). 2010. *Extreme Events in Nature and Society.* Berlin: Springer.

Anderson, Chris. 2008. *The Long Tail: Why the Future of Business is Selling Less of More.* New York: Hyperion.

Andriani, P. and B. McKelvey. 2009. From Gaussian to Paretian thinking: Causes and implications of power laws in organizations. *Organization Science* 20(6):1053–1071.

Anupindi, R., S. Chopra, S. D. Deshmukh, J. A. Van Mieghem, and E. Zemel. 2011. *Managing Business Process Flows.* Upper Saddle River, NJ: Prentice Hall.

Apgar, D. 2006. *Risk Intelligence: How to Live with Uncertainty.* Cambridge, MA: Harvard Business School Press.

Appleyard, B. 2012. Some like it very hot. *Intelligent Life*, May–June, 66–71.

Arthur, W. B. 1994. *Increasing Returns and Path Dependence in the Economy.* Ann Arbor: University of Michigan Press.

Arthur, W. B. 2009. *The Nature of Technology: What It Is and How It Evolves.* New York: Free Press.

Ashby, W. R. 1956. *Introduction to Cybernetics.* New York: John Wiley.

Bak, P. 1996. *How the Nature Works: The Science of Self-Organized Criticality.* New York: Copernicus Press.

Barks, C., J. Moyne, A. J. Arberry, and R. Nicholson. 1997. *The Essential Rumi.* Edison, NJ: Castle Books.

Barnes-Schuster D., Y. Bassok, and R. Anupindi. 2002. Coordination and flexibility in supply contracts with options. *Manufacturing and Service Operations Management* 4:171–207.

Barrett, F. J. 1998. Creativity and Improvisation in jazz and organizations: Implications for organizational learning. *Organization Science* 9(5):605–622.

Barrieu, P. and O. Scaillet. 2010. A primer on weather derivatives. In: Filar, Jerzy A. and Haurie, Alain (Eds.), *Uncertainty and Environmental Decision Making: A Handbook of Research and Best Practice.* Berlin: Springer, 155–176.

Bar-Yam, Y. 2004. Multiscale variety in complex systems. *Complexity,* 9(4):37–45.

Behar, R. 2008. China in Africa. *Fast Company*, June, 100–123.

Beinhocker, E. D. 2007. *The Origin of Wealth: Evolution, Complexity, and the Radical Remaking of Economics.* London: Random House.

Bell, D. E. 1985. Disappointment in decision making under uncertainty. *Operations Research* 33:1–27.

Ben-Haim, Y. 2006. *Info-Gap Theory: Decisions under Severe Uncertainty,* 2nd ed. London: Academic Press.

Berliner, P. F. 1994. *Thinking in Jazz: The Infinite Art of Improvisation.* Chicago: University of Chicago Press.

Bernstein, A. 2007. Gary Pisano: The thought leader interview. *Strategy+Business,* Summer, Issue 47. Available at www.strategy-business.com/media/file/sb47.07210.pdf (Accessed November 28, 2012.)

Bernstein, P. L. 1996. *Against the Gods: The Remarkable Story of Risk.* New York: John Wiley & Sons.

Bier, V. and M. A. Mosleh. 1990. The analysis of accident precursors and near misses: Implications for risk assessment and risk management. *Reliability Engineering and System Safety* 27:91–101.

Bier, V. M., Y. Y. Haimes, J. H. Lambert, N. C. Matalas, and R. Zimmerman. 1999. A survey of approached for assessing and managing the risk of extremes. *Risk Analysis* 19(1):83–94.

Birol, F. 2012. Energy and climate—Is our climate path already locked-in? Paper presented, İstanbul, Turkey, March 16. Regional and Global Climate Change: Physical Observations and Policy Choices.

Bloom, J. and S. Blair. 1997. *Islamic Arts.* London: Phaidon Press.

Böhmer, H., J. Powell, and S. Atılhan. 2008. *Nomads in Anatolia, Their Life and Their Textiles: Encounters with a Vanishing Culture.* Remhöb Verlag, Germany: Ganderkesee.

Bowden, A. 2010. The economic costs of maritime piracy, one earth future foundation. Working paper. Available at www.oceansbayandpiracy.org/sites/default/files/documents_old/the_economic_cost_of_piracy_full_report.pdf (Accessed June 25, 2012).

Bradenburger, A. and B. Nalebuff. 1996. *Co-Opetition.* New York: Doubleday.

Brindley, C. (Ed.). 2004. *Supply Chain Risk.* Burlington, VT: Ashgate Publishing.

Brockett, P., L. M. Wang, and C. Yang. 2005. Weather derivatives and weather risk management. *Risk Management and Insurance Review* 8:127–140.

Brown, B., M. Chui, and J. Manyika. 2011. Are you ready for the era of "big data"? *McKinsey Quarterly,* October, vol. 4, pp. 24–35.

Brunk, G. G. 2002. Why do societies collapse? A theory based on self-organized criticality. *Journal of Theoretical Politics* 14(2):195–230.

Brynjolfsson, E., L. M. Hitt, and H. H. Kim. 2011. Strength in numbers: How does data-driven decision making affect firm performance? Social Science Research Network (SSRN). Available at http://papers.ssrn.com/sol3/papers.cfm?abstract_id=1819486. (Accessed June 2, 2012.)

Buyck, C. 2009. Pegasus rides the LCC wave. Available at http://atwonline.com/it-distribution/article/pegasus-rides-lcc-wave-0309 (Accessed March 24, 2012).

Buzan, T. 2000. *Head First: 10 Ways to Tap into Your Natural Genius.* London: Thorsons, HarperCollins.

Cachon, G. 2003. Supply chain coordination with contracts. In: de Kok, A. G. and Graves, S. C. (Eds.), *Handbooks in Operations Research and Management Science: Supply Chain Management: Design, Coordination and Operation.* Amsterdam: Elsevier. pp. 229–239.

Çakanyıldırım, M. and Ç. Haksöz. 2012. *Global Perspectives: Turkey*. Lombard, IL: Council of Supply Chain Management Professionals.

Caldentey, R. and G. Bitran. 2003. An overview of pricing models for revenue management. *Manufacturing and Service Operations Management* 5:203–229.

Caldentey, R. and M. Haugh. 2006. Optimal control and hedging of operations in the presence of financial markets. *Mathematics of Operations Research* 31:285–304.

Can, Ş. 2002a. *Mesnevi Tercümesi.Birinci ve İkinci Cilt*. Ötüken Neşriyat, İstanbul, Turkey.

Can, Ş. 2002b. *Mesnevi Tercümesi.Üçüncü ve Dördüncü Cilt*. Ötüken Neşriyat, İstanbul, Turkey.

Cannon, M. D. and A. C. Edmondson. 2005. Failing to learn and learning to fail (intelligently): How great organizations put failure to work to innovate and improve. *Long Range Planning* 38(3):299–319.

Carmeli A. and J. Schaubroeck. 2008. Organization crisis preparedness: The importance of learning from failures. *Long Range Planning* 41:177–196.

Carpenter, S., B. Walker, J. M. Anderies, and N. Abel. 2001. From metaphor to measurement: Resilience of what to what? *Ecosystems* 4:765–781.

Carr, N. 2007. The ignorance of crowds. *Strategy and Business*, 47. Available at http://www.strategy-business.com/article/07204?pg=all (Accessed July 11, 2012).

Casti, J. L. 2010. *Mood Matters: From Rising Skirt Lengths to the Collapse of World Powers*. New York: Copernicus Books.

Casti, J. L. 2011. Four faces of tomorrow. OECD International Futures Project on Future Global Shocks. Available at http://www.oecd.org/dataoecd/57/60/46890038.pdf (Accessed June 17, 2012).

Casti, J. L. 2012. *X-Events: The Collapse of Everything*. New York: HarperCollins.

Chance, D. M. and R. Brooks. 2008. *An Introduction to Derivatives and Risk Management*. Mason, OH: Thomson South-Western.

Chen, F. and A. Federgruen. 2000. Mean–variance analysis of basic inventory models. Working paper, Graduate School of Business, Columbia University, New York.

Chen, F. Y. and C. A. Yano. 2010. Improving supply chain performance and managing risk under weather-related demand uncertainty. *Management Science* 56:1380–1397.

Chen, X., M. Sim, D. Simchi-Levi, and P. Sun. 2006. Risk aversion in inventory management. Working paper, University of Illinois at Urbana-Champaign.

Chernobai, A. S., S. T. Rachev, and F. J. Fabozzi. 2007. *Operational Risk: A Guide to Basel II Capital Requirements, Models, and Analysis*. Hoboken, NJ: John Wiley & Sons.

Chichilnisky, G. 1996. Catastrophe bundles can deal with unknown risk. *Best's Review*. Vol 96, Issue 10, pp. 44–48.

Chopra, S. and M. Sodhi. 2004. Managing risk to avoid supply-chain breakdown. *MIT Sloan Management Review* 46(1):53–61.

Clegg, E. 2002. How extremophiles thrive in impossible conditions: Lessons for business from nature's super-survivors. Available at http://www.visualinsight.net/extremeophile.html (Accessed August 5, 2012).

Cook, J. 1993. *The Book of Positive Quotations*. New York: Gramercy Book.

Cooperrider, D. L. and S. Srivastva. 1987. Appreciate inquiry in organizational life. *Research in Organizational Change and Development* 1:129–169.

Corbett, J. C. and K. Rajaram. 2006. A generalization of the inventory pooling effect to nonnormal dependent demand. *Manufacturing & Service Operations Management* 8(4):351–358.

Corcoran, W. R. 2004. Defining and analyzing precursors. In: Phimister, J. R., Bier, V. C., and Kunreuther, H. (Eds.), *Accident Precursor Analysis and Management: Reducing Technological Risk through Diligence.* Washington, DC: National Academies Press, 79–88.

Coutu, D. L. 2002. How resilience works. *Harvard Business Review*, May, 80(5), 46–55.

Crook, C. 2009. Complexity theory: Making sense of network effects. In: Kleindorfer, P. R. and Wind, J. (Eds.), *The Network Challenge: Network-Based Strategies and Competencies.* Upper Saddle River, NJ: Wharton School Publishing. pp. 207–223.

Crowther, K. G., Y. Y. Haimes, and G. Taub. 2007. Systemic valuation of strategic preparedness through application of the inoperability input-output model with lessons learned from Hurricane Katrina. *Risk Analysis* 27(5):1345–1364.

Cruz, M. and M. Pinedo. 2008. Total quality management and operational risk in the service industries. In Chen, Z., Raghavan, S. (Eds.), *Tutorials in Operations Research Informs:* Hanover, MD, 154–169.

Cruz, M. G. 2002. *Modeling, Measuring and Hedging Operational Risk.* Hoboken, NJ: John Wiley & Sons.

Dada, M., N. C. Petruzzi, and L. B. Schwarz. 2007. A newsvendor's procurement problem when suppliers are unreliable. *Manufacturing & Service Operations Management* 9(1):9–32.

Damodaran, A. 2007. *Strategic Risk Taking: A Framework for Risk Management.* Upper Saddle River, NJ: Wharton School Publishing.

Day, G. S. P. and J. H. Schoemaker. 2006. *Peripheral Vision: Detecting the Weak Signals that Will Make or Break Your Company.* Boston, MA: Harvard Business School Press.

De Bono, E. 1985. *Tactics: The Art and Science of Success.* London: Profile Books.

De Bono, E. 1991. *Opportunities.* London: Penguin Books.

De Bono, E. 2004. London: *How to Have a Beautiful Mind.* Tower, MN: Vermilion.

De Vany, A. W. and D. Walls. 1996. Bose-Einstein dynamics and adaptive contracting in the motion picture industry. *Economic Journal* 106:1493–1514.

Diamond, J. 2005. *Collapse: How Societies Choose to Fail or Succeed.* New York: Penguin Books.

Dillon, F. 2011. Don't break the chain. *The Irish Times*, November 25, p. 38.

Ding, Q., L. Dong, and P. Kouvelis. 2007. On the integration of production and financial hedging decisions in global markets. *Operations Research* 55(3):470–489.

Dong, L. and H. Liu. 2007. Equilibrium forward contracts on nonstorable commodities in the presence of market power. *Operations Research* 55:128–145.

Dörner, D. 1996. *The Logic of Failure: Recognizing and Avoiding Error in Complex Situations.* Cambridge, MA: Perseus Books.

Dye, R. 2008. The promise of prediction markets: A roundtable. *McKinsey Quarterly*. Issue 2, pp. 82–93.

Edmondson, A. C. 2011. Strategies for learning from failure. *Harvard Business Review,* April, 49–55.

Eeckhoudt, L., C. Gollier, and H. Schlesinger. 1995. The risk-averse (and prudent) newsboy. *Management Science* 41(5):786–794.

Elliott, M. R., P. R. Kleindorfer, J. J. DuBois, Y. Wang, and I. Rosenthal. 2008. Linking OII and RMP data: Does everyday safety prevent catastrophic loss? *International Journal of Risk Assessment and Management* 10(1–2):130–146.

Eppen, G. 1979. Effects of centralization on expected costs in a multilocation newsboy problem. *Management Science* 25(5):498–501.

Erdös, P. and A. Renyi. 1959. On random graphs. *Publicationes Mathematicae* 6:290–297.

Evans, D. 2012. *Risk Intelligence: Learning to Manage What We Do Not Know.* New York: Free Press.

Faroqhi, S. 1984. *Towns and Townsmen of Ottoman Anatolia: Trade, Crafts and Food Production in an Urban Setting, 1520–1650.* Cambridge, UK: Cambridge University Press.

Ferdows, K., J. A. Machuca, and M. Lewis. 2002. Zara. ECCH, The Case for Learning, Reference No. 603-002-1.

Findley, C. V. 2005. *The Turks in World History.* New York: Oxford University Press.

Fisher, M. and A. Raman. 1996. Reducing the cost of demand uncertainty through accurate response to early sales. *Operations Research* 44(1):87–99.

Florida, R. 2002. *The Rise of the Creative Class: And How It's Transforming Work, Leisure, Community, and Everyday Life.* New York: Basic Books.

Fung, V. K., W. K. Fung, and Y. Wind. 2008. *Competing in a Flat World: Building Enterprises for a Borderless World.* Philadelphia: Wharton School Publishing.

Funston, R. and S. Wagner. 2010. *Surviving and Thriving in Uncertainty: Creating the Risk Intelligent Enterprise.* New York: Wiley.

Gallego, G. and G. van Ryzin. 1994. Optimal dynamic pricing of inventories with stochastic demand over finite horizons. *Management Science* 40(8):999–1020.

Gardner, H. 1999. *Intelligence Reframed: Multiple Intelligences for the 21st Century.* New York: Basic Books.

Gaur, V. and S. Seshadri. 2005. Hedging inventory risk through market instruments. *Manufacturing & Service Operations Management* 7(2):103–120.

Gaur, V., S. Seshadri, and M. Subrahmanyam. 2011. Optimal timing of inventory decisions with price uncertainty. Working paper, Cornell University, Ithaca, NY.

Gawande, A. 2007. *Better: A Surgeon's Notes on Performance.* New York: Picador.

Giegengack, R. and Y. Bordeaux. 2009. Information networks in the history of life. In: Kleindorfer, P. R. and Wind, J. (Eds.), *The Network Challenge: Network-Based Strategies and Competencies,* p. 105–123. Upper Saddle River, NJ: Wharton School Publishing.

Gladwell, M. 2000. *The Tipping Point: How Little Things Can Make a Big Difference.* London: Abacus.

Goel, A. and G. J. Gutierrez. 2004. Integrating spot and futures commodity markets in the optimal procurement policy of an assembly-to-order manufacturer. Working paper, Department of Management, University of Texas, Austin.

Gomory, R. 1995. The known, the unknown, and the unknowable. *Scientific American,* 272. Issue 6, p. 120, June 1995.

Granovetter, M. 1973. The strength of weak ties. *American Journal of Sociology* 78(6):1360–1380.

Graves, S. C. and B. T. Tomlin. 2003. Process flexibility in supply chains. *Management Science* 49(7):907–919.

Gray, A. 2012. Disasters expose flaws in assumptions. *Financial Times,* Special Report on Risk Management-Supply Chain. March 20. Available at http://search.proquest.com/docview/931914981?accountid=13638. (Accessed November 29, 2012.)

Grove, A. S. 1999. *Only the Paranoid Survive: How to Exploit the Crisis Points that Challenge Every Company.* New York: Crown Business.

Gul, F. 1991. A theory of disappointment aversion. *Econometrica* 59:667–686.

Hagel, J. and J. S. Brown. 2005. *The Only Sustainable Edge: Why Business Strategy Depends on Productive Friction and Dynamic Specialization.* Boston, MA: Harvard Business School Press.

Haksöz, Ç. 2007. Quality management. *Wall Street Journal,* September 19, p. 12.

Hakšöz, Ç. 2008. An ounce of prevention is worth a pound of recalls. *Wall Street Journal*, August 6, p. 12.

Hakšöz, Ç. 2012. Supply chain risk perception: A cross-industry study in Turkey. Working paper, Sabancı University, Sabancı School of Management, İstanbul, Turkey.

Hakšöz, Ç. and A. Kadam. 2008. Supply risk in fragile contracts. *MIT Sloan Management Review* 49(2):7–8.

Hakšöz, Ç. and A. Kadam. 2009. Supply portfolio risk. *Journal of Operational Risk* 4(1):59–77.

Hakšöz, Ç. and B. Dağ. 2012. Pricing weather derivatives for Turkey with bootstrapping. Working paper, Sabancı School of Management, Sabancı University, İstanbul, Turkey.

Hakšöz, Ç. and D. D. Uşar. 2011. Silk Road supply chains: A historical perspective. In: Hakšöz, Ç., Seshadri, S., and Iyer, A. V. (Eds.), *Managing Supply Chains on the Silk Road: Strategy, Performance, and Risk*, p. 3–26. Boca Raton, FL: Taylor and Francis/CRC Press.

Hakšöz, Ç. and K. D. Şimşek. 2010. Modeling breach of contract risk through bundled options. *Journal of Operational Risk* 5(3):3–20.

Hakšöz, Ç. and S. Seshadri. 2007. Supply chain operations in the presence of a spot market: A review with discussion. *Journal of Operational Research Society* 58(11):1412–1429.

Hakšöz, Ç. and S. Seshadri. 2011. Integrated production and risk hedging with financial instruments. In: Kouvelis, P., Dong, L., Boyabatli, O., and Li, R. (Eds.), *Handbook of Integrated Risk Management in Global Supply Chain*. p. 157–196. New York: John Wiley & Sons.

Hakšöz, Ç., S. Seshadri, and A. V. Iyer (Eds.). 2011. *Managing Supply Chains on the Silk Road: Strategy, Performance, and Risk*. Boca Raton, FL: Taylor and Francis/CRC Press.

Hakšöz, Ç., K. Kandemir, and H. M. Camcı. 2012. Historical evolution of global supply chain risks. *Technical Report*, Sabancı School of Management, Sabancı University, İstanbul, Turkey.

Hammer, M. 2004. Deep change: How operational innovation can transform your company. *Harvard Business Review* 82:84–93.

Hämäläinen, R. and E. Saarinen. 2007. *Systems Intelligence in Leadership and Everyday Life*. Helsinki: Systems Analysis Laboratory, Helsinki University of Technology.

Hayes, R., G. Pisano, D. Upton, and S. Wheelwright. 2005. *Operations, Strategy, and Technology: Pursuing the Competitive Advantage*. Hoboken, NJ: John Wiley & Sons.

Heal, G. and H. Kunreuther. 2005. IDS models of airline security. *Journal of Conflict Resolution* 41:201–217.

Helbing, D. 2010. Systemic risks in society and economics. International Risk Governance Council (IRGC) Report. Available at http://irqc.gov/IMG/pdf/systemic_risks_Helbing2.pdf. (Accessed November 29, 2012.)

Hendricks, K. B. and V. R. Singhal. 2003. The effect of supply chain glitches on shareholder wealth. *Journal of Operations Management* 21(5):501–522.

Heracleous, L. and J. Wirtz. 2010. Singapore Airlines' balancing act. *Harvard Business Review*, Vol. 88, Issue 718, 145–149.

Hoening, C. 2000. *The Problem Solving Journey: Your Guide for Making Decisions and Getting Results*. Cambridge, MA: Perseus Publishing.

Hoffman, D. G. 2002. *Managing Operational Risk: 20 Firmwide Best Practice Strategies*. New York: John Wiley & Sons.

Holland, J. H. 1995. *Hidden Order: How Adaptation Builds Complexity*. New York: Basic Books.

Howe, N. and W. Strauss. 2007. The next 20 years: How customer and workforce attitudes will evolve. *Harvard Business Review*, July–August, vol. 85, issue 718, p. 41–52.

Hull, J. C. 2003. *Options, Futures, and Other Derivatives*. Englewood Cliffs, NJ: Prentice-Hall.

Isaka, S. 1997. Production woes at Toyota show price of cost-cutting. *Nikkei Weekly (Japan)*, February 10, Section: Major stories, p. 1.

Iyer, A., S. Seshadri, and R. Vasher. 2009. *Toyota Supply Chain Management: A Strategic Approach to the Principles of Toyota's Renowned System*. New York: McGraw-Hill.

Jakob, F. 1977. Evolution and tinkering. *Science* 196(4295):1161–1166.

Johansen, A., D. Sornette, and O. Ledoit. 1999. Predicting financial crashes using discrete scale invariance. *Journal of Risk* 1(4):5–32.

Johnson, S. 2010. *Where Good Ideas Come From: The Natural History of Innovation*. New York: Riverhead Books.

Jordan, W. C. and S. C. Graves. 1995. Principles on the benefits of manufacturing process flexibility. *Management Science*, 41(4):577–594.

Kambil, A. and V. Mahidhar. 2005. Disarming the value killers: A risk management study. *Deloitte Research*. Available at www.deloitte.com/assets/Deom-china/local/20assets/documents/Ch_CG_disarmingthevaluekillers_300807.pdf

Kamrad, B. and A. Siddique. 2004. Supply contracts, profit sharing, switching, and reaction: Options. *Management Science* 50:64–82.

Kauffman, S. A. 2000. *Investigations*. New York: Oxford University Press.

Kavanagh, M. 2012. Squeezing suppliers: Pragmatic approach defuses tensions in complex relationship. *Financial Times*, March 19. Available at http://search.proquest.com/docuview/929067559?accountid=136381. (Accessed November 29, 2012.)

Kay, J. 2011. *Obliquity: Why Our Goals Are Best Achieved Indirectly*. New York: Penguin Books.

Kelley, T. and J. Littman. 2001. *The Art of Innovation: Lessons in Creativity from IDEO, America's Leading Design Firm*. New York: Currency Books.

Kelly, K. 1998. *New Rules for the New Economy: 10 Radical Strategies for a Connected World*. New York: Penguin Books.

Khan, O. and G. Zsidisin. 2011. *Handbook for Supply Chain Risk Management: Case Studies, Effective Practices and Emerging Trends*. Fort Lauderdale, FL: J. Ross Publishing.

Kim, C. 2011. Toyota aims for quake-proof supply chain. *Reuters*. Available at http://www.reuters.com/article/2011/09/06/us-toyota-idUSTRE7852RF20110906 (Accessed October 12, 2011).

Kirchsteiger, C. 1997. Impact of accident precursors on risk estimates from accident databases. *Journal of Loss Prevention in the Process Industries* 10(3):159–167.

Kleindorfer, P. R. and G. Saad. 2005. Managing disruption risks in supply chains. *Production and Operations Management* 14(1):53–68.

Kleindorfer, P. R., M. R. Elliot, Y. Wang, and R. A. Lowe. 2004. Drivers of accident preparedness and safety: Evidence from the RMP Rule. *Journal of Hazardous Materials* 115:9–16.

Kleindorfer, P. R. and Y. Wind. 2009 (Eds.). *The Network Challenge: Network-Based Strategies and Competencies*. Upper Saddle River, NJ: Wharton School Publishing.

Kogut, B. and N. Kulatilaka. 1994. Operating flexibility, global manufacturing, and the option value of a multinational network. *Management Science* 40(1):123–139.

Kouvelis, P., K. Axarloglou, and V. Sinha. 2001. Exchange rate and the choice of ownership structure of production facilities. *Management Science* 47(8):1063–1080.

Kouvelis, P., L. Dong, O. Boyabatli, and R. Li. (Eds.). 2011. *Handbook of Integrated Risk Management in Global Supply Chains*. New York: John Wiley & Sons.

Kunreuther, H. 2009. The weakest link: Risk management strategies for dealing with interdependencies. In: Kleindorfer, P. R. and Wind, J. (Eds.), *The Network Challenge: Network-Based Strategies and Competencies*. Upper Saddle River, NJ: Wharton School Publishing.

Kunreuther, H. and P. Slovic. 1999. Coping with stigma: Challenges and opportunities. *Risk: Health, Safety & Environment* 10:269–280, p. 383–398.

La Porte, T. R. 1996. High reliability organizations: Unlikely, demanding and at risk. *Journal of Contingencies and Crisis Management* 4(2):60–71.

Lamarre, E., M. Pergler, and G. Vainberg. 2009. Reducing risk in your manufacturing footprint. *McKinsey Quarterly*, April. Available at www.mckinseyquarterly.com/Reducing_risk_in_your_manufacturing_footprint_23501. (Accessed June 2, 2012.)

Lampel, J. and Z. Shapira. 2001. Judgmental errors, interactive norms, and the difficulty of detecting strategic surprises. *Organization Science* 12(5):599–611.

Langer, E. J. 1989. *Mindfulness*. Cambridge, MA: Da Capo Press.

Lederer, P. J. and R. S. Nambimadom. 1998. Airline network design. *Operations Research* 46(6):785–804.

Lee, H. 2004. The triple-A supply chain. *Harvard Business Review*, October, 102–112.

Levin, S. A. 1999. *Fragile Dominion: Complexity and Commons*. Cambridge, MA: Perseus Publishing.

Levinthal, D. 1997. Adaptation on rugged landscapes. *Management Science* 43:934–950.

Li, C. L. and P. Kouvelis. 1999. Flexible and risk-sharing supply contracts under price uncertainty. *Management Science* 45:1378–1398.

Liker, J. K. 2004. *The Toyota Way: 14 Management Principles from the World's Greatest Manufacturer*. New York: McGraw-Hill.

Lindblom, C. E. 1959. The science of "muddling through." *Public Administration Review* 19(2):79–88.

Linnenluecke, M. and A. Griffiths. 2010. Beyond adaptation: Resilience for business in light of climate change and weather extremes. *Business Society* 49(3):477–511.

Loewenstein, G. F., E. U. Weber, C. K. Hsee, and N. Welch. 2001. Risk as feelings. *Psychological Bulletin* 127(2):267–286.

Machowiak, W. 2008. The critical approach to risk assessment: Making the risk impact analyzable. Paper presented at *2nd European Risk Conference*, September 11–12, 2008. Bocconi University, Milan.

Madina, J. J. 2008a. *Brain Rules: 12 Principles for Surviving and Thriving at Work, Home, and School*. Seattle, WA: Pear Press.

Madina, J. J. 2008b. The science of thinking smarter. *Harvard Business Review*, May, vol. 86(5):51–54.

Mahoney, P. G. 2005. Contract remedies and options pricing. *Journal of Legal Studies* 24(1):139–163.

Makridakis, S., R. M. Hogarth, and A. Gaba. 2009. Forecasting and uncertainty in the economic and business world. *International Journal of Forecasting* 25:794–812.

Makridakis, S. N. and N. Taleb. 2009. Living in a world of low levels of predictability. *International Journal of Forecasting* 25:840–844.

Malik, Y., A. Niemeyer, and B. Ruwadi. 2011. Building the supply chain of the future. *McKinsey Quarterly*, January p. 62–71.

Mandelbrot, B. B. 1983. *Fractal Geometry of Nature*. San Francisco, CA: W.H. Freeman.

Mandelbrot, B. B. and R. L. Hudson. 2005. *The (Mis)Behavior of Markets: A Fractal View of Risk, Ruin and Reward*. London: Profile Books.

Manisaligil, A. and Ç. Haksöz. 2012. Risk, resilience, and organizational improvisation. Working paper, Sabancı School of Management, Sabancı University, İstanbul, Turkey.

March, J. G., L. S. Sproull, and M. Tamuz. 1991. Learning from samples of one or fewer. *Organization Science* 2(1):1–13.

Martinez-de-Albéniz, V. and D. Simchi-Levi. 2005. A portfolio approach to procurement contracts. *Production and Operations Management* 14(1):90–114.

Massey, C. and G. Wu. 2005. Detecting regime shifts: The causes of under- and overreaction. *Management Science* 51(6):932–947.

Matthews, R. G. 2008. Corporate news: Steelmakers squeeze suppliers—as weak demand weighs on raw-material prices, mills look to jettison contracts. *Wall Street Journal*, November 18, vol. 252, Issue 119, p. B2.

May, R. M., S. A. Levin, and G. Sugihara. 2008. Ecology for bankers. *Nature* 451:893–895.

Maynard, M. 2010. Toyota official says recall may not fully solve safety problem. *New York Times*. Available at http://www.nytimes.com/2010/02/24/business/global/24toyota.html?_r=1&ref (Accessed February 24, 2010).

McCann, K. S. 2000. The diversity-stability debate. *Nature* 405:228–233.

McDonald, R. L. and D. R. Siegel. 1985. Investment and the valuation of firms when there is an option to shut down. *International Economic Review* 26(2):331–349.

McKay, G. 2004. *The Encyclopedia of Animals: A Complete Visual Guide*. San Francisco, CA: Fog City Press.

McKelvey, B. and P. Andriani. 2010. Avoiding extreme risk before it occurs: A complexity science approach to incubation. *Risk Management* 12(1):54–82.

McNeil, A., R. Frey, and P. Embrechts. 2005. *Quantitative Risk Management: Concepts, Techniques, Tools*. Princeton, NJ: Princeton University Press.

Mecham, M. 2009. Unions need to apply. *Aviation Week and Space Technology* 171(22):41.

Michalko, M. 2003. A theory about genius. Available at http://creativethinking.net/DA01_ATheoryAboutGenius.htm?Entry=Good (Accessed June 18, 2012).

Michel, J., Y. K. Shen, A. P. Aiden, A. Veres, M. K. Gray, and W. Brockman. 2010. The Google books team. Quantitative Analysis of Culture Using Millions of Digitized Books. *Science* 331:176–182.

Michel-Kerjan,E. and F. Morlaye. 2007. Extreme events, global warming, and insurance-linked securities: How to trigger the tipping point. Working Paper # 2007-09-13, Wharton School, Philadelphia.

Mokyr, J. 1990. *The Lever of Riches: Technological Creativity and Economic Progress*. New York: Oxford University Press.

Muallimoğlu, N. 1998. *Turkish Delights: A Treasury of Proverbs and Folk Sayings*. İstanbul, Turkey: Milli Eğitim Bakanlığı Yayınları, Düşünce Eserleri Dizisi.

Mürmann, A. and U. Oktem. 2002. The near-miss management of operational risk. *Journal of Risk Finance* 4(1):25–36.

Nadin, M. 2006. Anticipating extreme events. In: Albeverio, S., Jentsch, V., and Kantz, H. (Eds.), *Extreme Events in Nature and Society*, p. 21–45. Berlin: Springer.

Nagali, V., J. Hwang, D. Sanghera, M. Gaskins, M. Pridgen, T. Thurston, P. Mackenroth, D. Branvold, P. Scholler, and G. Shoemaker. 2008. Procurement risk management (PRM) at Hewlett-Packard company. *Interfaces* 38(1):51–60.

Narasimhan, R., A. Nair, A. D. Griffith, J. S. Arlbjørn, and E. Bendoly. 2009. Lock-in situations in supply chains: A social exchange theoretic study of sourcing arrangements in buyer–supplier relationships. *Journal of Operations Management* 27(5):379–385.

Netessine, S., G. Dobson, and R. A. Shumsky. 2002. Flexible service capacity: Optimal investment and the effect of demand correlation. *Operations Research* 50(2): 375–388.

Nishiguchi, T. and A. Beaudet. 1998. The Toyota group and the Aisin fire. *MIT Sloan Management Review*, Fall, vol. 40, Issue 1, 49–59.

Norberg, J. G. and S. Cumming (Eds.). 2008. *Complexity Theory for a Sustainable Future.* New York: Columbia University Press.

Obrien, J. 2002. Pounds 35M goodwill demand sets the alarm bells ringing. *Birmingham Post*, May 21, p. 21.

Ohmae, K. 1982. *The Mind of the Strategist: The Art of Japanese Business.* New York: McGraw-Hill.

Ohno, T. 1988. *Toyota Production System: Beyond Large-Scale Production.* Portland, OR: Productivity Press.

Oktem, U., R. Wong, and R. C. Oktem. 2010. Near-miss management: Managing the bottom of the risk pyramid. *Risk and Regulation*, July 2010. 12–13.

Osorio, I. M., G. Frei, D. Sornette, J. Milton, and Y. C. Lai. 2009. Epileptic seizures: Quakes of the brain. Available at http://arxiv.org/abs/0712.3929v1. (Accessed August 7, 2012.)

Page, S. E. 2006. Path dependence. *Quarterly Journal of Political Science* 1:86–115.

Page, S. E. 2007. *The Difference: How the Power of Diversity Creates Better Groups, Firms, Schools, and Societies.* Princeton, NJ: Princeton University Press.

Page, S. E. 2011. *Diversity and Complexity.* Princeton, NJ: Princeton University Press.

Panjer, H. H. 2006. *Operational Risk Modeling Analytics.* Hoboken, NJ: John Wiley & Sons.

Pascale, R. T., J. Sternin, and M. Sternin. 2010. *The Power of Positive Deviance: How Unlikely Innovators Solve the World's Toughest Problems.* Boston, MA: Harvard Business Press.

Paté-Cornell, M. E. 1986. Warning systems in risk management. *Risk Analysis* 6(2):223–234.

Paté-Cornell, M. E. 1996. Global risk management. *Journal of Risk and Uncertainty* 12:239–255.

Paté-Cornell, M. E. 2002. Fusion of intelligence information: A Bayesian approach. *Risk Analysis* 22(3):445–454.

Paté-Cornell, M. E. 2004. On signals, response, and risk mitigation: A probabilistic approach to the detection and analysis of precursors. In: Phimister, J. R., Bier, V. C., and Kunreuther, H. (Eds.), *Accident Precursor Analysis and Management: Reducing Technological Risk through Diligence.* Washington, DC: National Academies Press, 45–59.

Perrett, B. 2011. Broken chain. *Aviation Week and Space Technology* 173(10):20.

Perrow, C. 1999. *Normal Accidents: Living with High-Risk Technologies.* Princeton, NJ: Princeton University Press.

Perry, L. T. 1991. Strategic improvising: How to formulate and implement competitive strategies in concert. *Organizational Dynamics* 19(4):51–64.

Persidis, A. 1998. Extremophiles. *Nature Biotechnology* 16:593–594.

Pettit, T. J., J. Fiksel, and K. L. Croxton. 2010. Ensuring supply chain resilience: Development of a conceptual framework. *Journal of Business Logistics* 31(1):1–20.

Phimister, J. R., U. Oktem, P. R. Kleindorfer, and H. Kunreuther. 2003. Near-miss incident management in the chemical process industry. *Risk Analysis* 23(3): 445–459.

Phimister, J. R., V. M. Bier, and H. Kunreuther (Eds.). 2004. *Accident Precursor Analysis and Management: Reducing Technological Risk through Diligence.* Washington, DC: National Academies Press.

Pinedo, M. 2010. (Ed.) *Operational Control in Asset Management: Processes and Costs.* Copenhagen: Simcorp StrategyLab.

Pinedo, M., S. Seshadri, and E. Zemel. 2002. The Ford-Firestone case. New York: New York University Stern School of Business.

Popova, M. 2012. Elevator groupthink: A psychology experiment in conformity, 1962. Available at http://www.brainpickings.org/index.php/2012/01/13/asch-elevator-experiment/(Accessed July 11, 2012).

Prahalad, C. K. and M. S. Krishnan. 2008. *The New Age of Innovation: Driving Co-Created Value through Global Networks.* New York: McGraw-Hill.

Regester, M. 2000. Prevention is better than cure. In: Mars, G. and Weir, D. (Eds.), *Risk Management.* Aldershot, UK: Ashgate/Dartmouth, 143–153.

Ridderstrale, J. and K. Nordstrom. 2007. *Funky Business Forever: How to Enjoy Capitalism.* Harlow, UK: Financial Times/Prentice Hall.

Rijpma, J. A. 1997. Complexity, tight-coupling and reliability: Connecting normal accidents theory and high reliability theory. *Journal of Contingencies and Crisis Management* 5(1):15–23.

Rilke, R. M. 1947. Letters of Rainer Maria Rilke. vol. 2, 1910–1926. Translated by J. B. Greene and M. D. H. Norton. www.Norton and Company. New York, NY.

Ritchie, B. and C. Brindley. 2007. An emergent framework for supply chain risk management and performance management. *Journal of the Operational Research Society* 58:1398–1411.

Ritchken, P. H. and C. S. Tapiero. 1986. Contingent claims contracting for purchasing decisions in inventory management. *Operations Research* 34:864–870.

Rivkin, J. W. 2000. Imitation of complex strategies. *Management Science* 46(6):824–844.

Roberto, M. A. 2002. Lessons from Everest: The interaction of cognitive bias, psychological safety, and system complexity. *California Management Review* 45(1):136–158.

Rock, D. 2009. Managing with the brain in mind. *Strategy+Business* 56, Autumn. www.strategy-business.com/media/file/sb56_09306.pdf (Accessed November 29, 2012.)

Roe, E. P. and R. Schulman. 2008. *High Reliability Management: Operating on the Edge.* Stanford, CA: Stanford Business Books.

Rosenthal, I., P. Kleindorfer, H. Kunreuther, E. Michel-Kerjan, and P. Schmeidler. 2004. Lessons learned from chemical accidents and incidents, Discussion document, Wharton Risk Management and Decision Processes Center of the University of Pennsylvania. Philadelphia, PA.

Ross, B. 2011. The importance of risk type in selecting appropriate analytic approaches and management strategies. Paper presented at the *Annual Meeting for the Society for Risk Analysis,* Charleston, SC., December 4–7, 2011.

Sak, H. and Ç. Haksöz. 2011. A copula-based simulation model for supply portfolio risk. *Journal of Operational Risk* 6(3):15–38.

Sandmo, A. 1971. On the theory of the competitive firm under price uncertainty. *American Economic Review* 61:65–73.

Santayana, G. 1925. Dialogues in limbo. London: Constable and Co. Available at http://archive.org/details/dialoguesinlimbo014715mbp) (Accessed June 4, 2012).

Scheffer, M. 2009. *Critical Transitions in Nature and Society.* Princeton, NJ: Princeton University Press.

Scheffer, M., J. Bascompte, W. A. Brock, V. Brovkin, S. R. Carpenter, V. Dakos, H. Held, E. H. van Nes, M. Rietkerk, and G. Sugihara. 2009. Early-warning signals for critical transitions. *Nature* 461(3):53–59.

Scheffer, M., S. Carpenter, J. A. Foley, C. Folke, and B. Walker. 2001. Catastrophic shifts in ecosystems. *Nature* 413:591–596.

Schelling, T. 1978. *Micromotives and Macrobehavior*. New York: Norton.

Schwalbe, M. 2012. The 40-30-30 rule: Why risk is worth it. Available at http://the99percent.com/tips/6103/The-40-30-30-Rule-Why-Risk-Is-Worth-It (Accessed June 17, 2012).

Seifert, R., W. Thonemann, and W. Hausman. 2004. Optimal procurement strategies for online spot markets. *European Journal of Operational Research* 152:781–799.

Senge, P. M. 1990. *The Fifth Discipline: The Art and Science of the Learning Organization*. New York: Doubleday.

Sezer, A. D. and Ç. Haksöz. 2012. Optimal decision rules for product recalls. *Mathematics of Operations Research*. 37(3): 399–418.

Sheffi, Y. 2007. *The Resilient Enterprise: Overcoming Vulnerability for Competitive Advantage*. Cambridge, MA: MIT Press.

Shekerjian, D. 1990. *Uncommon Genius: How Great Ideas are Born*. New York: Penguin Books.

Simon, H. 1962. Architecture of complexity. *Proceedings of the American Philosophical Society* 106(6), p. 467–482.

Simon, H. 1986. How managers express their creativity. *Across the Board* 23(3):11–17.

Simonton, D. K. 1999. *Origins of Genius: Darwinian Perspectives on Creativity*. New York: Oxford University Press.

Slovic, P. 1987. Perception of risk. *Science* 236:280–285.

Slovic, P., B. Fischoff, and B. S. Lichtenstein. 1982. Facts versus fears: Understanding perceived risk. In: Kahneman, D., Slovic, P., and Tversky, A. (Eds.), *Judgment under Uncertainty: Heuristics and Biases*, 463–493. Cambridge University Press, New York, NY.

Slovic, P. 1993. Perceptions of Environmental Hazards: Psychological Perspectives. In Gärling, I. and Golledge, R. G. (Eds.), *Behavior and Environment: Psychological and Geographical Approaches*. Amsterdam, Netherlands. Elsevier, 223–248.

Snavely, B. 2006. Visteon pays $14.9 million award in contract dispute. *Automative News*, October 23. Vol. 81, Issue 6226, p. 16.

Sodhi, M. and C. Tang. 2012. *Managing Supply Chain Risk*. New York: Springer.

Son, L. K. and N. Kornell. 2010. The virtues of ignorance. *Behavioral Processes* 83:207–212.

Sornette, D. 2002a. Economies of scale in innovations with block-busters. *Quantitative Finance* 2:224–227.

Sornette, D. 2002b. Predictability of catastrophic events: Material rupture, earthquakes, turbulence, financial crashes, and human birth. *PNAS* 99(1):522–2529.

Sornette, D. 2003. Critical market crashes. *Physics Reports* 378:1–98.

Sornette, D. 2006a. *Critical Phenomena in Natural Science: Chaos, Fractals, Self-Organization and Disorder: Concepts and Tools*. Berlin: Springer-Verlag.

Sornette, D. 2006b. Endogenous versus exogenous origins of crises. In: Albeverio, S., Jentsch, V., and Kantz, H. (Eds.), *Extreme Events in Nature and Society*. 95–119. Berlin: Springer.

Sornette, D. 2008. Nurturing breakthroughs: Lessons from complexity. *Journal of Economic Interaction and Coordination* 3:165–181.

Sornette, D. 2009. Dragon-kings, black swans and the prediction of crises. *International Journal of Terraspace Science and Engineering*. Available at http://arxiv.org/abs/0907.4290v1. (Accessed August 9, 2012.)

Spinler, S., A. Huchzermeier, and P. Kleindorfer. 2003. Risk hedging via options contracts for physical delivery. *OR Spectrum* 25(3):379–395.

Stapleton, O., L. Stadtler, and L. N. Van Wassenhove. 2011. Private-humanitarian supply chain partnerships on the Silk Road. In: Haksöz, Ç., Seshadri, S., and Iyer, A. I. (Eds.), *Managing Supply Chains on the Silk Road: Strategy, Performance, and Risk*. p. 217–238. Boca Raton, FL: Taylor and Francis/CRC Press.

Stone, E. R., J. F. Yates, and A. M. Parker. 1994. Risk communication: Absolute versus relative expressions of low-probability risks. *Organizational Behavior and Human Decision Processes* 60:387–408.

Surowiecki, J. 2004. *The Wisdom of Crowds: Why the Many are Smarter than the Few and How Collective Wisdom Shapes Business, Economics, Societies, and Nations*. New York: Doubleday.

Sümer, F. 1999. *Oğuzlar (Türkmenler)*. İstanbul, Turkey: Türk Dünyası Araştırmaları Vakfı.

Sutton, R. I. 2002. *Weird Ideas that Work: How to Build a Creative Company*. New York: Free Press.

Taleb, N. N. 2007. *The Black Swan: The Impact of the Highly Improbable*. New York: Random House.

Tang, C. and B. Tomlin. 2008. The power of flexibility for mitigating supply chain risk. *International Journal of Production Economics* 116(1):12–27.

Tang, C. S. 2006. Perspectives in supply chain risk management. *International Journal of Production Economics* 103(2):451–488.

Tapiero, C. 2004. *Risk and Financial Management: Mathematical and Computational Concepts*. London: Wiley.

Tapiero, C. S. 2005. Value at risk and inventory control. *European Journal of Operational Research* 163(3):769–775.

Tevelson, R. M., J. Ross, and P. Paranikas. 2007. *A Key to Smart Sourcing: Hedging against Risk*. Boston: Boston Consulting Group.

The Economist. 2012. The roar of the crowd. May 26. p. 77–79.

Thomke, S. H. 2003. *Experimentation Matters: Unlocking the Potential of New Technologies for Innovation*. Boston, MA: Harvard Business School Press.

Thomke, S. H. 2010. Design thinking and innovation at Apple. Harvard Business School Case Study No. 609-066, Boston, MA.

Toffler, A. and H. Toffler. 2006. *Revolutionary Wealth: How It Will Be Created and How It Will Change Our Lives*. New York: Knopf.

Tomlin, B. 2009. Impact of supply learning when suppliers are unreliable. *Manufacturing & Service Operations Management* 11(2):192–209.

Tripp, R. T. 1970. *The International Thesaurus of Quotations*. New York: Thomas Y. Crowell Company.

Tversky, A. and D. Kahneman. 1974. Judgment under uncertainty: Heuristics and biases. *Science* 185(4157):1124–1131.

Van Mieghem, J. A. 1998. Investment strategies for flexible resources. *Management Science* 44(8):1071–1078.

Van Mieghem, J. A. 2003. Capacity management, investment and hedging: Review and recent developments. *Manufacturing & Service Operations Management* 5(4):269–302.

Van Mieghem, J. A. and N. Rudi. 2002. Newsvendor networks: Dynamic inventory management and capacity investment with discretionary pooling. *Manufacturing & Service Operations Management* 4(4):313–335.

Vanderhaeghe, A. and S. de Treville. 2003. How to fail at flexibility. *Supply Chain Forum: An International Journal* 4(1):67–73.

Yakubovich, V. and R. Burg. 2009. Missing the forest for the trees: Network-based HR strategies. In: Kleindorfer, P. R. and Wind, J. (Eds.), *The Network Challenge: Network-Based Strategies and Competencies.* p. 335–352. Upper Saddle River, NJ: Wharton School Publishing.

Yi, W. and V. M. Bier. 1998. An application of copulas to accident precursor analysis. *Management Science* 44(12):257–270.

Wagner, S. M., C. Bode, and P. Koziol. 2009. Supplier default dependencies: Empirical evidence from the automotive industry. *European Journal of Operational Research* 199(1):150–161.

Waldrop, M. M. 1992. *Complexity: The Emerging Science at the Edge of Order and Chaos.* New York: Penguin Books.

Walker, B. and D. Salt. 2006. *Resilience Thinking: Sustaining Ecosystems and People in a Changing World.* Washington, DC: Island Press.

World Economic Forum. (WEF). 2011. Global risks report, 6th ed. Available at http://reports.weforum.org/global-risks-2011/. (Accessed June 2, 2012.)

World Economic Forum. (WEF). 2012a. New models for addressing supply chain and transport risk report. Available at http://www.weforum.org/reports/new-models-addressing-supply-chain-and-transport-risk. (Accessed June 2, 2012.)

World Economic Forum (WEF). 2012b. Global Risks Landscape. Available at www.reports.weforum.org/global-risks-2012/. (Accessed November 29, 2012.)

Weick, K. E. 1998. Improvisation as a mindset for organizational analysis. *Organization Science* 9(5):543–555.

Weick, K. E. and K. M. Sutcliffe. 2007. *Managing the Unexpected: Resilient Performance in an Age of Uncertainty.* San Francisco, CA: Jossey-Bass.

Weick, K. E., K. M. Sutcliffe, and D. Obstfeld. 1999. Organizing for high reliability: Processes for collective mindfulness. In: Sutton, R. S. and Staw, B. M. (Eds.), *Research in Organizational Behavior,* vol. 1. Stanford, CA: JAI Press, 81–123.

Wind, Y., C. Crook, and R. Gunther. 2006. *The Power of Impossible Thinking: Transform the Business of Your Life and the Life of Your Business,.* Upper Saddle River, NJ: Prentice Hall.

Wu, D. J., P. R. Kleindorfer, and J. E. Zhang. 2002. Optimal bidding and contracting strategies for capital-intensive goods. *European Journal of Operational Research* 137:657–676.

Zander, R. S. and B. Zander. 2000. *The Art of Possibility: Transforming Professional and Personal Life.* New York: Penguin Books.

Zsidisin, G. A. and B. Ritchie. 2010. *Supply Chain Risk: A Handbook of Assessment, Management, and Performance.* New York: Springer.

Zsidisin, G. A., S. M. Wagner, S. A. Melynk, G. L. Ragatz, and L. A. Burns. 2008. Supply risk perceptions and practices: an exploratory comparison of German and US supply management professionals. *International Journal of Technology, Policy and Management* 8(4):401–419.

Index

C

D

E

N

O

P

Q

R

T

U

V

W

For Product Safety Concerns and Information please contact our EU representative GPSR@taylorandfrancis.com
Taylor & Francis Verlag GmbH, Kaufingerstraße 24, 80331 München, Germany

www.ingramcontent.com/pod-product-compliance
Ingram Content Group UK Ltd.
Pitfield, Milton Keynes, MK11 3LW, UK
UKHW021831240425
457818UK00006B/164

* 9 7 8 1 4 6 6 5 0 4 4 7 9 *